AF653098

Trends and Issues of Livestock Production Systems

Trends and Issues of Livestock Production Systems

Editor

Dr. Bibek Ghosh

2007

Gene-Tech Books

New Delhi - 110 002

2007, © Publisher

Information contained in this work has been published by Gene-Tech Books and has been obtained by its author(s)/editor(s) from sources believed to be reliable and are correct to the best of their knowledge. However, the publisher and its author(s) make no representation of warranties with respect of accuracy or completeness of the contents of this book, and shall in no event be liable for any errors, omissions or damages arising out of use of this information and specifically disclaim any implied warranties or merchantability or fitness for any particular purpose.

All rights reserved. Including the right to translate or to reproduce this book or parts thereof except for brief quotations in critical reviews.

ISBN 81-89729-66-7

Published by : **GENE-TECH BOOKS**
4762-63/23, Ansari Road, Darya Ganj,
NEW DELHI - 110 002
Phone: 41562849
e-mail: genetechbooks@yahoo.co.in

Printed at : **Tarun Offset Printers**
Delhi

PRINTED IN INDIA

Preface

Exponentially expanding human populations raises the demand for all agricultural products and increases the stress on the resource base used for their production. Rapidly growing demand for livestock products worldwide is brought about by human population pressure, growing income and urbanisation. Land use and human population pressures are leading to intensification and expansion in many livestock production systems. In addition, expansion of cropping into drier areas is forcing pastoral livestock production systems to relocate into still more arid lands. As a result of these changes new pressures on the environment are developing or could emerge and, therefore, should be of concern.

The scale and nature of the interaction between livestock production and the environment has been the subject of much conjecture, all of which has lacked a technical basis for making informed policy decisions and devising technical intervention programmes. However, it is increasingly clear that livestock-environment linkages should be seen in the context of human, economic and political aspects as well as natural resource utilisation.

Characteristics of animal agriculture systems have been developed in response to agro-ecological opportunities and demand for livestock commodities. In many cases, a fully sustainable equilibrium has been established. Furthermore, in many of these environmentally balanced systems, the livestock element is interwoven with crop production. Animal manure is often the essential element in maintaining soil fertility. In other cases, such as the semi-nomadic pastoral systems of many of the world's natural grassland regions, environmentally stable balances of human society, animal population and vegetative biomass have existed for centuries.

Livestock make an important contribution to most economies. Livestock produce food, provide security, enhance crop production, generate cash incomes for rural and urban populations, provide fuel and transport, and produce value added goods which can have multiplier effects and create a need for services. Furthermore, livestock diversify production and income, provide year-round employment, and spread risk. Livestock also form a major capital reserve of farming households. Because of livestock's contribution to societies, human and economic pressures can direct livestock production in ways detrimental to the environment.

This book addresses the trends and issues in livestock production systems and attempts to asses more objectively the role of livestock in environmentally sustainable agriculture. It can help organisations and professionals in finding orientation in the world's livestock sector that is becoming increasingly complex.

Editor

Contents

1

Livestock Production Systems

Livestock are domesticated animals intentionally reared in an agricultural setting to make produce such as food or fibre, or for their labour. Livestock include pigs, cattle, goats, deer, sheep, yaks and poultry. The type of livestock reared varies worldwide and depends on factors such as climate, consumer demand, native animals, local traditions, and land type. Livestock may be raised for subsistence food or for profit. Raising animals is an important component of modern agriculture. It has been practiced in many societies, since the transition to farming from hunter-gather lifestyles.

Animal-rearing has its origins in the transition of societies to settled farming communities rather than hunter-gather lifestyles. Animals are 'domesticated' when their breeding and living conditions are controlled by humans. Over time, the collective behaviour, life cycle, and physiology of livestock have changed radically. Many modern farm animals are unsuited to life in the wild. Goats, sheep, and pig were domesticated around 8000BC in Asia. The earliest evidence of horse domestication dates to around 4000 BC.

Purpose of Animal Rearing

'Livestock' are defined, in part, by their end purpose as the production of food or fiber, or labour. The economic value of livestock includes:

— *Meat*: The production of a useful form of dietary protein and energy.

— *Dairy products*: Mammalian livestock can be used as a source of milk, which can in turn easily be processed into other dairy products such as yogurt, cheese, butter, ice cream, kefir, and kumiss. Using livestock for this purpose can often yield several times the food energy of slaughtering the animal outright.

— *Fiber:* Livestock produce a range of fiber/textiles. For example, sheep and goats produce wool; cows, deer, and sheep can make leather; and bones, hooves and horns of livestock can be used.

— *Fertiliser*: Manure can be spread on fields to increase crop yields. This is an important reason why historically, plant and animal domestication have been intimately linked. Manure is also used to make plaster for walls and floors and can be used as a fuel for fires. The blood and bone of animals are also used as fertiliser.

— *Labour*: Animals such as horses, donkey, and yaks can be used for mechanical energy. Prior to steam power livestock were the only available source of non-human labour. They are still used for this purpose in many places of the world, including ploughing fields, transporting goods, and military functions.

— *Land Management*: The grazing of livestock is sometimes used as a way to control weeds and undergrowth. For example, in areas prone to wild fires, goats and sheep are set to graze on dry scrub which removes combustible material and reduces the risk of fires.

During the history of animal husbandry many secondary products have arisen in an attempt to increase carcass utilisation and reduce waste. For example, animal offal and non-edible parts may be transformed into products such as pet food and fertiliser. In the past such waste products were sometimes also fed to livestock as well. However, intra-species recycling poses a disease risk, threatening animal and even human health. Due primarily to BSE (mad cow disease), feeding animal scraps to animals has been banned in many countries, at least in regards to ruminants.

Classification of Livestock Systems

Livestock production systems are considered to be a subset of farming systems. A review of the literature revealed that most farming systems classifications are not backed by quantitative criteria, which would enable cases to be clearly allocated to one class. These classifications are closer to typologies. No attempts at developing a classification of world livestock systems by using quantitative statistical methodologies could be located in the literature. This probably relates to the lack of appropriate data sets for such approaches on a global scale.

The classification of livestock systems are generally based on criteria such as integration with crops, relation to land and agro-ecological zone. Some forms of livestock systems based on these criteria are:

— *Solely Livestock Production Systems (L)*: Livestock systems in which more than 90 percent of dry matter fed to animals comes from rangelands, pastures, annual forages and purchased feeds and less than 10 percent of the total value of production comes from non-livestock farming activities.

— *Landless Livestock Production Systems (LL)*: Subset of the solely livestock production systems in which less than 10 percent of the dry matter fed to animals is farm-produced and in which annual average stocking rates are above ten livestock units (LU) per hectare of agricultural land.

— *Grassland-based Systems (LG)*: Subset of solely livestock production systems in which more than 10 percent of the dry matter fed to animals is farm-produced and in which annual average stocking rates are less than ten LU per hectare of agricultural land.

— *Mixed-farming Systems (M)*: Livestock systems in which more than 10 percent of the dry matter fed to animals comes from crop by-products or stubble or more than 10 percent of the total value of production comes from non-livestock farming activities.

— *Rain-fed mixed-farming systems (MR)*: A subset of the mixed systems in which more than 90 percent of the value of non-livestock farm production comes from rain-fed land use.

— *Irrigated mixed-farming systems (MI)*: A subset of the mixed systems in which more than 10 percent of the value of non-livestock farm production comes from irrigated land use.

The features of each system, with quantitative estimates of the magnitude of the resources involved in each system defined, together with the major outputs and a set of productivity indices are described below.

Landless Livestock Production Systems (LL)

The developed countries dominate the picture of landless intensive production with more than half of total meat production. Asia is already contributing some 20 percent and eastern Europe 15 percent, with the latter recently in sharp decline.

Landless Monogastric Production System (LLM)

This system is defined by the use of monogastric species, mainly chickens and pigs, where feed is introduced from outside the farm, thus separating decisions concerning feed use from those of feed production, and particularly of manure utilisation on fields to produce feed and/or cash crops. This system is therefore open in terms of nutrient flow.

Landless monogastric systems are found predominantly in OECD member countries with 52 percent of the total landless pork production and 58 percent of the landless poultry production globally. In the case of pig production, Asia is second, with 31 percent of the world total. For poultry, Central and South America follow, with 15 percent. To a large extent, this geographic distribution is determined by markets and consumption patterns in addition to levels of urbanisation.

In Southeast and eastern Asia, this system is especially important. As much as 96 percent of the total pig-meat production in Asia occurs in China, Viet Nam and Indonesia. China, Thailand and Malaysia produce 84 percent of poultry meat. This is associated with fast economic growth and urbanisation. The demand for monogastric meat is expected to grow from two- to fivefold between 1987 and 2006 from a base of 31 million tonnes, and a three- to tenfold increase is expected in the demand for eggs from 9 million tonnes.

The prerequisites for development into large-scale vertically integrated production include the use of appropriate breeds and strains, feed quantity and quality, housing and disease control, as well as assured markets both at home and abroad. Landless poultry and pig production systems account for the majority of the output in developed countries and their share is rapidly increasing in developing countries given their high supply elasticity in the short term.

The system is typically competing with traditional land-based production systems for shares in the urban markets. It must be kept in mind that poultry and pork are close substitutes for beef and mutton, thus also interacting with the ruminant systems. In a broader sense, the demand for cereals created by these systems is also competing for land resources with land-based ruminant systems.

Landless Ruminant Production System (LLR)

This production system is defined by the use of ruminant species, principally cattle, where feed is mainly introduced from outside the farm system. Landless ruminant production systems are highly concentrated in only a few regions of the world. In the case of cattle, they are almost exclusively found in eastern Europe and the CIS and in a few OECD member countries. Landless sheep production systems are only found in western Asia and northern Africa.

Typical cases are large-scale feedlots in the United States and in eastern Europe and the CIS, and veal production in parts of the KU. Intensive dairy operations in the same regions are more land-based because of the need to feed palatable fodder, which cannot be transpo-rted economically over long distances. In Asia, the intensive ruminant systems are typically found in buffalo and dairy cattle production units/colonies in India and Pakistan.

The LLR system is based almost exclusively on high-producing, specialised breeds and their crosses, which, nevertheless, have not been bred specifically for performance under landless conditions. With regard to milk production, the Holstein-Friesian breed is clearly the most important, and for beef production, English breeds predominate in the United States, while the large European dual-purpose breeds provide animals for fattening. The abundance of rangeland in the United States has led to the specialised production of calves from beef breeds for feedlot operations, while under European conditions these animals are a joint product together with milk, mainly from mixed systems.

The LLR system is highly capital-intensive, leading to substantial economies of scale. It is also feed-intensive and labour-extensive. Key efficiency parameters are daily weight gains and feed conversion, basically reflecting the efficient use of capital invested in infrastructure or in the form of lean animals and feeds.

Weight gains are usually in the range of 1 to 1.5 kg per day, and feed conversion rates are about 8 to 10 kg of grains per kilogram of weight gain. This system is closely linked to land-based systems that normally provide the young stock for landless systems. This constitutes an important difference from landless monogastric systems, in which replacement stock is produced within the same vertically integrated system.

Driven by population growth, the landless ruminant system is expected to continue to grow slowly in North America and southern Asia. On the other hand, it is expected to decline in the EU as production becomes more extensive in response to policies reducing agricultural support. In eastern Europe and the CIS, its importance is also declining and ruminant production in that part of the world is shifting to land-based and smaller-scale systems.

A growing market for grain-fed beef exists in Japan and the newly industrialised countries of Asia. The growth rate of this market will depend mainly on the evolution of the international price of cereals and the increased per caput incomes. This market will in part be supplied domestically and through imports from the United States, Canada, Australia and possibly South America.

Grassland-based Livestock Production Systems (LG)

The grassland-based systems are seen in different parts of the world. Central and South America and the developed countries dominate the picture in terms of meat production, together accounting for more than three-quarters of the world's production.

Temperate Zones and Tropical Highlands Grassland-based System (LGT)

In these areas, the grazing system is constrained by low temperatures. In the temperate zones, there are one or two months of mean temperatures, corrected to sea level, to below 5°C, whereas in the tropical highlands daily mean temperatures during the growing period are in the range of 5° to 20°C.

Locations in the tropical highlands comprise parts of the highlands of South America and eastern Africa, and in temperate zones they include southern Australia, New Zealand and parts of the United States, China and Mongolia.

Typical cases are Mongolia's steppe system, New Zealand's dairy and sheep enterprises, dairy systems close to Bogota, Colombia, and the South American camelid and sheep grazing systems in the altiplano of Peru and Bolivia. Extensive grazing systems are also found in parts of northwestern Pakistan involving sheep for mutton and wool and transhumant sheep on degraded high-altitude pasture in Nepal. Further cases are reported for Chinese merino wool sheep on communal grazing in Jilin Province and sheep ranching on grass-clover pastures of intensive animal production enterprises in the state of Oregon in the United States.

The regions in which the LGT system predominates have a combined human population of 190 million, which represents only 3.5 percent of the world total, and almost half of this population lives in Asia. In OECD member countries, far fewer people use the LGT system, but they control more land and cattle per inhabitant than in the other regions.

Under the LGT system, product use varies widely, ranging from export-oriented New Zealand farmers, to South American farmers mainly producing for the domestic market, to Asian and African smallholders concerned with local markets and their own subsistence. Market forces and environmental concerns are curbing the potential for intensification of this system. As a result, their global market share is declining vis-à-vis that of other production systems.

Since the LGT system is found mostly in marginal areas, its production potential in global terms is relatively low. In developing countries it tends to form a subsistence basis for certain groups of the population and its future role is seen more in providing employment for these groups than in making a major contribution to output and economic development.

In developed countries, frequently with production surpluses, the production from these systems is declining in relation to other values and uses assigned to these land resources, such as recreational value, value as a wildlife and biological diversity reserve and the contribution to water conservation.

Humid and Subhumid Tropics and Subtropics Grassland-based System (LGH)

The LGH system is defined as a grazing system found in regions with more than 180 days of growing period. It tends to be concentrated more in the subhumid zone, particularly in regions where access to markets or, for agronomic reasons, crop production is limited. By definition, only very limited cropping is considered for subsistence.

The LGH system is found mostly in the tropical and subtropical lowlands of South America: in the llanos of Colombia and Venezuela as well as in the cerrados of Brazil. Dual-purpose milk-beef systems in the Mexican lowlands and estancias in Argentina are also typical cases of this system. In the African setting, many of the potentially suitable land resources are not used as a result of trypanosomiasis constraining livestock production. Outside Latin America, this system is important only in Australia because of its ample land resources in relation to its population.

Worldwide, the LGH system comprises about 190 million head of cattle, an important proportion of which are of the zebu breeds. In the subhumid and humid regions, cattle are clearly the dominant species, and in very high rainfall areas, such as the Amazon River delta and some parts of Queensland, Australia, buffaloes are also ranched. African hair sheep and dwarf goats are usually kept for local consumption only. In the subtropics, wool sheep are an important component of the system, for example, in Argentina, Uruguay, South Africa and Australia.

The LGH system produces approximately 6 million tonnes of beef and veal and 11 million tonnes of cow milk worldwide. By far the most important geographic region is Central and South America. The system is predominantly market-oriented.

Globally, 6 percent of the world's population (330 million people) lives in areas where the LGH production system predominates. Its importance in terms of sustaining the livelihood of rural populations is expected to decline as interaction with crop cultivation turns it into a mixed system. In rain-forest regions, efforts are being made to incorporate perennial tree crops, frequently into silvo-pastoral systems. In the savannahs, this system is being converted into a mixed-farming system by including annual crops, such as maize, soybeans and sorghum.

Improvements in road infrastructure and new technologies to allow the establishment of pastures with commercially worthwhile nurse crops are making the ley farming systems, involving rotations of crops and pasture, a potentially attractive pathway into mixed-farming systems.

Arid and semi-arid tropics and subtropics grassland-based system (LGA). The LGA system is defined as a land-based system in tropical and subtropical regions with a growing period of less than 180 days, and where grazing ruminants is the dominant form of land use.

This system is found under two contrasting socioeconomic frameworks: in sub-Saharan Africa and the Near East and North Africa regions, where it constitutes a traditional way of subsistence for an important part of the population, and in Australia, parts of western United States and southern Africa, where private enterprises utilise publicly or privately owned range resources for ranching purposes.

While in western Asia, northern Africa and sub-Saharan Africa, the LGA system is important for the livelihood of large sectors of the rural population, in developed countries it is extremely labour-extensive.

In sub-Saharan Africa, agropastoralism is the most important interface between livestock production and other agricultural production. In other regions these systems are interrelated with other livestock production systems that have access to better-quality feed and are closer to markets. In low-income countries without an export market, incentives to produce quality beef are weak. This, in turn, limits the attractiveness of livestock production stratification.

Globally, new roles are emerging for these rangelands, besides that of producing ruminant animal products. In developing countries, the pressure to expand crop production is increasing population pressure on the remaining rangelands. In developed countries, the utilisation of these rangelands for animal production has often been subsidised through very low prices for grazing permits and public investments in irrigation.

Rain-fed Mixed-farming Systems (MR)

The mixed-farming systems are distributed all over the world. Sub-Saharan Africa, West Asia and North Africa, and Central and South America are relatively unimportant in terms of meat production, whereas developed countries and Asia together contribute about 70 percent of the total meat production from mixed-farming systems.

Temperate Zones and Tropical Highlands Rain-fed System (MRT)

This system is defined as a combination of rain-fed crop and livestock farming in temperate or tropical highland areas, in which crops contribute at least 10 percent of the value of total farm output.

The MRT system is found in two contrasting agro-ecozones of the world: it is the dominant system in most of North America, Europe and northeastern Asia, basically covering large strips of land north of the 30° northern latitude parallel, and it is found in the tropical highlands of eastern Africa (Ethiopia, Kenya, Uganda, Burundi, Rwanda) and in the Andean region of Latin America (Ecuador, Mexico).

The main common feature of these two regions is that low temperatures during all or part of the year limit and determine vegetation that is quite distinct from that found in tropical environments.

In the course of the development process, production technology in temperate locations evolved to higher specialisation, more use of external inputs and more open systems. This has resulted in increasing negative externalities of these systems for the environment.

In most tropical MRT systems, production is less intensive, with livestock performing a series of functions in mixed systems: a continuous flow of cash income; means to concentrate nutrients for crops through manure; fuel; animal traction; a cash reserve for emergencies; and as a buffer to risks in crop production.

Globally, the MRT system is the most important source of animal products, providing 39 percent of the beef and veal production, 24 percent of the mutton production and 63 percent of the cow milk produced.

In developed countries, growing environmental concern, reduced protection of domestic production and increased international trade have led to a stagnation or even a reduction in output levels. In eastern Europe and the CIS, mixed systems will replace inefficient large-scale landless systems as these economies open up to the markets.

In Asia, this system has the potential for increased production and better use of feeds for intensified ruminant production, associated with reallocation of land-use patterns at the farm level.

In the highlands of sub-Saharan Africa and Latin America, production increases must come from a further intensification of crop-livestock land-use systems. The highlands should favour small-scale mechanisation because of the high population density, continuous cultivation and relatively heavy soils.

Humid and Subhumid Tropics and Subtropics Rain-fed System (MRH)

In the humid and subhumid regions of the tropics and subtropics, livestock production is based on mixed-farming systems. Given the range of socioeconomic conditions and soils and climates involved, this livestock system is very heterogeneous in its composition. It is found in all tropical regions of the world, mainly in developing countries. Parts of the southern United States are the only significantly developed regions included in this system. Typical cases are smallholder rice-buffalo systems in Southeast Asia or soybean-maize-pasture large-scale commercial operations in the Brazilian cerrados.

This system includes regions with especially difficult climatic conditions for livestock. Adaptation of highly productive temperate breeds to these challenges has been notably poor. In many parts of Africa, trypanosomiasis constitutes an additional constraint to these systems.

Particularly in African and Asian smallholder systems the local breeds are still widely used. In Latin America, Bos taurus cattle, sheep and goats were introduced some four centuries ago. Bos indicus cattle were introduced a few decades ago and have now replaced the earlier introduced cattle breeds in tropical areas.

In the African and Asian MRH system, the multiple roles of livestock have prevailed, particularly animal traction and manure. In Central and South America, this system caters to large domestic markets and, particularly in the case of Brazil, it is also linked to export markets. Under smallholder conditions, milk tends to be a more important output than meat.

The MRH system applies to approximately 14 percent of the global population. This ratio is particularly high in sub-Saharan Africa, where 41 percent of the region's population is associated with the system, and in Central and South America, where it is 35 percent. The system is replacing grazing systems in Africa and Latin America. In Africa, the process is mainly driven by population growth, and, in Central and South America, by economic development and technological innovations.

The main challenge in sub-Saharan Africa is finding ways to increase productivity. It is generally acknowledged that the biological potentials of mixed systems will be the key to productivity increases, and the expectation is that purchased feed inputs will be replaced by nutrients cycled within the system. In Asia, increased crop production—and hence feed availability is an important way of intensifying and maximising animal productivity, however, the prospects for increasing cultivation area are limited.

In Latin America, low population density, a high degree of urbanisation and relatively high per caput incomes have resulted in farming systems that are generally more oriented towards livestock production. In the tropical rain-forest regions, high resource-consuming systems were established, in some cases driven by policies and in other cases by poverty. Many of the policies that promoted wasteful utilisation of these resources have been stopped in the process of structural adjustment.

Arid and Semi-arid Tropics and Subtropics Rain-fed System (MRA)

The MRA system is a mixed-farming system in tropical and subtropical regions with a vegetation growth period of less than 180 days. The main restriction of this system is the low primary productivity of the land resulting from low rainfall. The more severe the constraint, the less important crops become in the system and the more livestock take over as a primary income and subsistence source.

This system is important in the West Asia and North Africa region in parts of the Sahel (Burkina Faso, Nigeria), in large parts of India and in northeastern Thailand and eastern Indonesia, and less important in Central and South America. Given the low intensity of the system and the multiple purposes of livestock, the introduction of improved breeds has been limited. Globally, 11 percent of cattle and 14 percent of sheep and goats are found in this system.

Small ruminants are particularly important in West Asia and North Africa under the MRA system. As is the case in other largely smallholder systems, livestock have a range of simultaneous roles in this system, including animal traction, production of manure and use as a cash reserve, in addition to the production of meat and milk. Fuelwood is often scarce as a result of deforestation and range degradation, leading to the ever-increasing role of animals as providers of manure for fuel, in addition to means of transport. While this system supports larger populations than any other grazing system, only 10 percent of the world's population is related to this system. Fifty-one percent of the population involved is in Asia, mainly India, and 24 percent is in the West Asia and North Africa region.

There is close interaction with the LGA system. With increasing population pressure, the LGA system tends to evolve into mixed systems, mainly MRA, because of the greater caloric efficiency of cropping vis-à-vis ruminant production when land becomes scarce. The outlook for this system is relatively similar to that for the LGA system. The resource base puts a clear ceiling on agricultural intensification. Low and variable response to inputs makes their use financially risky. Population growth in this setting is contributing to the over-exploitation of the natural resource base, as traditional property rights cannot cope with the growing demands. Alternative development strategies and the reduction of population pressure on the resource base are key elements for the sustainable development of these regions.

Irrigated Mixed-farming Systems

Temperate zones and tropical highlands mixed system (MIT). This system belongs to the group of land-based mixed systems of temperate and tropical highland regions. It is found particularly in the Mediterranean region (Portugal, Italy, Greece, Albania, Bulgaria) and in the Far East (the Democratic People's Republic of Korea, the Republic of Korea, Japan and parts of China), where plant growth is limited both by low temperatures in the cold season and reduced moisture availability during the vegetation period. The system's importance in tropical highlands is negligible.

Meat, milk and wool, the main outputs of this system, are mainly produced for the market. Manure is an issue only where animals are stabled, at least for certain periods of the day or the year.

About 10 percent of the global population lives in regions where this system is dominant. A large share belongs to developed countries with relatively high income levels and where agricultural trade is important. This system is clearly associated with very intensive agriculture in temperate regions with a high population density. With the outcome of the General Agreement on Tariffs and Trade (GATT) negotiations, it can be expected that these systems will be less and less viable, having to compete with more efficient rain-fed systems producing the same commodities.

The expansion of international trade, and particularly the incorporation of southern European countries into the KU, has led to an increase in intensive production systems for off-season vegetables and fruits on the best irrigated land. The integration with livestock has been reduced, with ruminant grazing systems declining in absolute terms and being concentrated on the marginal sites.

Humid and subhumid tropics and subtropics mixed system (MIH). This is a mixed system in tropical and subtropical regions with growing seasons of more than 180 days, and in which the irrigation of crops is significant.

The MIH system is particularly important in Asia. High population densities require intensive crop production, and the irrigation of rice makes it possible to obtain more than two crops per year, even under conditions of very seasonal rainfall, substantially reducing yield variability as compared with the yield of upland rice or other rain-fed crops. In the past, animal production has been closely linked to the animal traction issue.

MIH systems throughout the world produce 13 million tonnes of pork (18 percent of global production), more than any other land-based tropical system. Among the tropical and subtropical systems, the MIH system is the one related to the largest population group, 990 million people, 97 percent of which are in Asia. Increasing labour productivity and relative affluence of farmers in this system are reflected in the more frequent use of tractors for cultivation.

Manure is recycled on the fields. Ducks do well in this environment, as do pigs and poultry, which thrive on abundant crop residues. In addition, since large ruminants, such as buffaloes and, to a lesser extent, cattle, have little association with draught and transport, their output will be increased milk and meat for the market.

Animals and intensive crop production in this ecological zone are an illustration of a successful and sustainable agricultural production system. Competition for urban markets for livestock products is the main form of interaction with the landless monogastric system, both domestically and globally, through international trade.

Arid and Semi-arid Tropics and Subtropics Mixed System (MIA)

This is a mixed system of arid and semiarid regions, in which irrigation makes year-round intensive crop production feasible. It is found in the Near East, South Asia, North Africa, western United States and Mexico.

Typical cases are alfalfa/maize-based intensive dairy systems in California, Israel and Mexico; small-scale buffalo milk production in Pakistan; and animal traction-based cash-crop production in Egypt and Afghanistan. Cattle and buffaloes for milk and animal traction are the main ruminant resources, although sheep and goats are important where marginal rangelands are available in addition to irrigated land.

In the MIA system, pigs are kept only in the Far East; they are virtually non-existent in West Asia and North Africa, largely for cultural reasons (Islamic and Jewish religions). The main introduced breeds are dairy cattle to supply milk to large urban centres. Under good management conditions, intensive dairy schemes have been quite successful in hot but dry environments. Some of the world's highest lactation yields are achieved in the MIA system in Israel and California. The traditional smallholder MIA system in Asia relies heavily on buffaloes for milk production.

In the traditional MIA system, irrigated crop production is the main source of income, with livestock playing a very secondary role. This is generally reflected in rather extensive management of livestock enterprises. The MIA system is predominant in regions that are home to over 750 million people, two-thirds of them in Asia and one-third in West Asia and North Africa. The main interaction with other systems occurs through the international market, particularly for milk and dairy products.

The MIA system makes an important contribution to food availability and employment in semi-arid and arid regions. Irrigation allows increased fodder production as a by-product or part of crop rotation, which reduces the feed deficit. The improved feed base and utilisation promotes intensification and commercialisation of livestock production, especially in areas with good market access.

Livestock Development Strategies

Livestock make a major, although largely underestimated, contribution to rural development in developing countries. They produce food, enhance crop production and provide additional economic goods and services as well as cash income. The inclusion of livestock diversifies and increases total farm production and income, provides year-round employment and disperses risk. Sales of livestock products provide funds for purchasing crop inputs and for financing farm investments. Livestock often form the major capital reserve of farming households and, in general, enhance the economic viability and sustainability of a farming system.

Despite these positive contributions to agriculture and economic development, many formal livestock projects have failed to meet their objectives, with the result that donors are becoming more and more reluctant to support such developments. Furthermore, animal production is increasingly being viewed far more critically:

— intensive production systems, particularly in industrialised countries, are seen as a major source of pollution;

— increasing ruminant numbers in developing countries are being associated with the degradation of the rangelands and soil erosion;

— livestock development is said to favour the richer segments of society — both producers and consumers — rather than the most vulnerable;

— livestock are thought to compete directly with humans for cereal grains.

Such controversies and the inherent complexity of livestock production impose constraints that have to be addressed and which pose particular challenges not normally faced by the agricultural planner. Yet it is the very complexity of animal production systems that also offers some of the greatest opportunities for development. Livestock, because of their linkages with the overall farming system, make valuable entry points for wider agricultural development programmes. To exploit these opportunities, an integrated approach that combines both technical and institutional interventions is required.

In many countries, the difficulties associated with increasing sustainable animal production are exacerbated by limited public-sector investment and weak, ineffective support services. Programmes and projects are often poorly designed and inadequately targeted, leading to the inefficient and fragmented allocation of scarce development resources. Policies related to the livestock sector are often incoherent with ill-defined goals and with little or no assessment of their likely impact.

The lack of consistent, integrated strategies that focus limited resources on identified and attainable goals remains a major constraint to livestock development. The situation is further complicated in that livestock, especially cattle, represent wealth and status and, as a consequence, are disproportionately owned by policy-makers who have a clear vested interest - an interest that is not necessarily beneficial to livestock development in general

Planning Perspective

For the agricultural planner, difficulties are related not only to the complexity of livestock production systems but also to an inability to understand how these systems function—this is primarily a problem of quantification and comprehension. One consequence is that development opportunities are often ignored, particularly the potential for using livestock as a catalyst to drive agricultural development. An understanding of the production factors and processes that affect animal production is a prerequisite for livestock development.

Production Factors

Livestock. Animals themselves are the major resource, but their mobility makes them a difficult resource to quantify, especially under the extensive management systems that predominate in much of the developing world. While this may create problems for the government statistician and the tax collector, it is unique in that it allows for the exploitation of feed resources for which there are no alternative productive uses, such as extensive arid, semi-arid and cold savannahs, crop residues and agricultural fallow.

Capital. Livestock ownership is more skewed than ownership of or access to land, and, as a consequence, livestock development, especially if it concerns larger and more expensive species such as cattle, tends to produce benefits of low equity. Many non-livestock owners are often precluded from livestock development because of lack of capital or credit. Yet, in many farming systems, livestock constitute the main, if not the only, capital reserve of farming households and, importantly, one that can be readily realised. As such, livestock serve as a strategic reserve that adds stability to the overall fanning system. In this respect, a mix of species increases stability; for example, cattle represent a long-term investment whereas sheep, goats and poultry are primarily shorter-term investments and sources of petty cash.

Feed. Whereas animals can substitute for capital, purchased feeds can substitute for land, creating "landless" production systems, where land is no longer a production factor. These are not primary production systems in the strictest sense, but more industrial and almost exclusively demand-driven. Conversely, extensive pastoral systems depend exclusively on forage, the availability of which is both seasonal and highly dependent on natural factors, most notably rainfall. Such systems are essentially resource-driven and less responsive to price changes.

Many animal feeds also have alternative uses, either for human consumption or for industrial use. These competing demands are determined by price and availability of the commodity, its use and the value of the end-product. In this respect, the more intensive, demand-driven production systems are an important alternative use of these commodities, and, as such, feed-conversion efficiency becomes an important productivity indicator and management objective.

Land. Escalating ruminant (large and small) populations, agricultural encroachment and decline in traditional authority have put an increasing strain on "open access" feed resources - particularly extensive grazing areas leading, in extreme cases, to irreversible degradation. The conflict between communal ownership of land and private ownership of livestock - the classic "tragedy of the commons" - has resulted in a disequilibrium that continues to threaten the ecological stability of many of these fragile environments. However, an increasing body of evidence suggests that these rangeland ecosystems have adapted and are more resistant to heavy stocking than previously thought, and that opportunistic range management often is an efficient and ecologically sound resource use. For the livestock planner, these are particularly complex and sensitive issues that must be addressed. Technical options are limited, and certainly solutions cannot be found without cognizance of the wider institutional and social contexts.

Access to water, especially in the extensive rangeland systems, is a fundamental prerequisite for livestock production. Historically, water has governed both access to and Use of these resources and has provided a brake on exploitation. Major environmental issues have recently arisen over the provision of new water sources, however, particularly perennial boreholes.

Labour. Labour is the other important resource for which livestock production has a specific requirement. Typically, livestock production is more labour-intensive and less seasonal than crop production, and can utilize family labour that usually has a low opportunity cost. The distribution of labour, responsibilities and benefits tends to be favourable for women, especially with the smaller animal species, which they themselves may own. Since women are largely responsible for the day-to-day management of the household, any development programme must take their time availability into account.

Production Process

Livestock production tends to be more complex than crop production. Production cycles, although seasonally influenced, are less pronounced with livestock. Some species have very short reproductive cycles (rabbits, poultry), while those of others (large ruminants) are far longer. In more intensive production systems, seasonal influences can be offset by modifying the environment through improved nutrition and artificially controlled light and temperature.

Unlike intensive production systems, which produce a single product, many of the production systems common to the developing world produce a range of commodities. These may include a mixture of consumables and services that provide cash, subsistence and inputs (draught power and manure) into the farm enterprise. Since livestock so often have a pivotal role in the overall farm system, any constraint imposed on livestock may also restrict the system as a whole.

Animals also have an important asset function. Many smallholders, particularly in mixed-farming systems, prefer the flow products (milk, draught, manure) rather than the end-products (meat, hides and skins) since selling their animals for slaughter entails the permanent loss of flow products. Only in larger herds or flocks can meat offtake be regarded as a flow product. Flow products generate a regular cash income, unlike end-products or crop revenues, and the importance of regular cash flow, however small, is often underestimated in many development efforts, especially within the smallholder sector.

Given the perishable nature of animal products, development beyond household consumption requires marketing and processing facilities that may not be readily available. Milk, for example, not only needs a continuous outlet, but also transport, storage and processing facilities. This interdependency between animal production and the links in the marketing chain increases as systems intensify.

The close relationship between domesticated animals and humans evident throughout history continues, and many societies have strong socio-cultural values attached to their animals. In most cases, these values reflect specific economic attributes and have important implications for livestock development.

Risk is another important factor in livestock production. Extensive livestock production systems are exposed to particularly high risks as a result of natural factors such as drought and disease, whereas in mixed-farming systems livestock help to reduce the overall risks to the enterprise. As production systems intensify, the production risks from natural causes decrease and are replaced by economic risks, such as price fluctuations and taxes. Producers can mitigate risk through diversification (mixed farming and mixed species), flexibility (choice of stocking rate) and increased productivity. The design of livestock development strategies must take these factors into account through risk analysis.

Consumption

On the consumption side, livestock products have characteristics that require specie-1 attention in the planning process. Meat, milk and eggs are all perishable products that need refrigeration and careful handling if they are to remain fit for human consumption. Such processes impose certain infrastructural requirements that may be beyond the resources of a given country or even a region. This is a major contributing factor to the small amount of milk products marketed in sub-Saharan Africa, for example.

Typically, livestock products also have a high elasticity of demand but a low elasticity of supply, particularly in land-based smallholder production. In many countries the demand for animal products increases rapidly with expanding urbanisation (and changes in consumption patterns), growth in per caput income and a growing population. Because of this demand pattern it has been argued that livestock development tends to favour the higher-income sectors of society - an isolated view, yet one that has deterred potential donors - but does not adequately take account of benefits on the supply side.

All edible livestock products have high nutritive value. They are especially rich in protein, with a favourable composition of amino acids. Milk availability is one of the major contributors to the alleviation of malnutrition among the more vulnerable sections of society, most notably children. The excessive consumption of animal products, which is now recognised as a serious health hazard, is primarily a concern in affluent societies, while in developing countries consumption levels remain low. Indeed, animal fat provides a valuable contribution to the daily energy requirements in these countries.

Livestock-sector Planning Approach

The FAO Animal Production and Health Division (AGA) is developing a methodology to provide a rational basis for livestock-sector planning that will assist governments in determining their own policies and priorities. The principles described apply equally to a particular segment, species or geographical region of the subsector-milk production, animal health or the subhumid zone, for example — and to the subsector as a whole. The chosen approach can be described as interdisciplinary, as it combines the biological, social and economic sciences, dynamic, in that it uses scenarios to anticipate future development, and systems-oriented.

In general, most livestock development strategies have similar aims, namely, to:

- — conserve the natural resource base;
- — raise productivity through better utilisation of available resources: capital (animals), land and labour;
- — expand production where there is a sufficient demand and resources can be utilised at reasonable cost to the environment;
- — optimise the allocation of development resources through rational administration and management.

A strategy for livestock development has a number of technical and socio-economic dimensions that need to be addressed if it is to have any relevance or acceptance. To make the planning process simple and transparent, the approach adopted by AGA follows the logical sequence of:

— *Diagnosis*. An analysis of the current performance of the livestock subsector, its potentials and constraints.

— *Prescription*. A creative phase in which programmes and policy options are formulated and designed to address identified constraints and potentials.

— *Prognosis*. A comparative (with and without) assessment of the likely impact and implications of the prescribed programmes and policy options.

The analytical process usually proceeds from the bottom up, that is, from the lowest production unit to its effects on the national economy, and contains an ex ante evaluation of ongoing as well as previous programmes and policy interventions. The process is interactive in that the results of any comparative assessment may require modifications to either the design or further analysis. The final result is the presentation of a series of development options along with a corresponding set of recommended actions and policies required if the options are to be implemented. Implicit in the success of any such exercise is a high degree of national "ownership" resulting from the maximum involvement of nationals in the complete sequence through such elements as rapid rural appraisals (RRAs), national consultancies and workshops and expert working groups.

With regard to livestock policy analysis and formulation, the concept of induced innovation is applied. According to this concept, the choice of technology is induced by a relative abundance of production factors, such as those introduced above, for example, in the case of livestock production. For a number of the effects that livestock production can have, such as those on the environment and on human health, no markets exist.

These effects are referred to as market failures, or externalities, which can be both positive and negative. Where these externalities exist, such as in the case of waste disposal from intensive pork and poultry production, incentives or regulations are part of policies that aim to reflect more accurately the implicit value of a non-marketed commodity, such as human health or the environment. Incentives and regulations affect factor use as they infringe on factor prices and, thus, on technological choice.

Diagnosis Phase

Livestock need to be placed within the overall context of the agricultural sector and the national economy as a whole. The first step is usually an inventory of the natural and economic resources available for livestock production: land availability and use (by agro-ecological zone); feed (crop residues, agro-industrial by-products, cereals, forages, dodders); animals (breeds, etc.); water; labour; capital; public infrastructure; and support services (marketing, input supply). Existing production methods and productivity, processing, storage and marketing all need to be assessed in relation to the available resources.

In many developing countries, traditional and commercial production systems exist alongside one another with little or no linkage between them. Industrial systems of poultry, pig and, to a lesser extent, dairy production are demand-driven and responsive to changing market conditions, due, in part, to their short productive cycles and a wide and varying range (in price and availability) of potential feeds. Conversely, traditional ruminant production systems are inelastic in supply and sometimes even respond inversely to price increases, as was evident in the negative beef supply response observed in Latin America and in southern Africa.

A number of macroeconomic policies also have direct and indirect impacts on livestock production. Pricing policies - often through the manipulation of exchange rates - wages and interest rates determine the costs and returns of livestock production. Fiscal policies govern government expenditures on, for example, rural infrastructures and producer or consumer subsidies.

Monetary policy determines the availability and cost of agricultural credit, while trade policies (tariffs, quotas and export subsidies) regulate the import (produce and raw material) and export of livestock products and animal feeds. Institutional structures have a bearing on a wide range of support services that affect livestock production, such as marketing (marketing boards), quality and hygiene standards, extension and research, and agricultural education. The balance between private and public ownership of livestock-related services is becoming a critical issue, especially as extended public services prove to be increasingly inefficient and costly to run. These macroeconomic and sectoral policies need to be analysed as to their relevance to livestock production; incentives and/or disincentives to livestock production as they affect relative abundancy of production factors; and adequacy in view of likely future developments.

In addition to these broad economic policies, countries usually pursue more specific sectoral policies that may or may not be interrelated with the overall policy framework. These policies usually refer to issues such as: land tenure; input supplies and pricing (feed industry, veterinary supplies, genetic resources); labour supply and wages; specific livestock credits; marketing of livestock products; product processing facilities; and capacity and performance of extension, training and research institutions. On a microlevel, policies may be analysed using econometric methods for farm household data, integrated farm household models, mathematical programming or a policy analysis matrix.

Livestock and the Environment

Production systems vary considerably in their impact on the environment. Most beneficial effects are found in mixed farming systems, whereas landless systems are facing a range of environmental problems, mostly related to waste and resource use. Pastoral systems are mostly environmentally balanced, but this equilibrium is endangered as demographic pressure persists. Within production systems, production intensity is a major factor determining the nature of livestock-environment interactions.

Within identified farming systems, farm households can be defined according to their available (or lack of) resources, production patterns, productivity and income. Such analysis is usually derived from diagnostic surveys such as RRAs and farm management budgeting (FMB). The breakdown of a livestock subsector into various production systems is justified given the differences that exist in available resources and applied technologies, as well as in primary products and their uses (subsistence or market-oriented), making it easier to deal with typologies.

It is at this level where most development efforts are concentrated, and in the majority of countries this usually entails a breakdown of the livestock subsector into production systems broadly based on agro-ecological zones and socio-economic patterns. Livestock production systems are basically a subset of farming systems, although their analysis must comprise the whole farm household.

Available feedstuffs may be classified by their physical properties (dry matter, texture, fibre content), chemical properties (protein, energy) or origin (on-farm, purchased, communal). Inadequate feed intake and/or unbalanced diets are usually the primary production constraint. In addition, feed is also a major cost component, especially in the more intensive production systems. Available feed resources need to be carefully assessed, therefore, with estimated feed requirements in mind.

Livestock resources consist of various species; their breeds and genetic potential, herd composition and ownership are the predominant determinants of a production system. These factors change as production systems become more an integral part of the economy. With increasing commercialisation, productivity and demand, there is a move from predominantly resource-driven systems to those primarily demand-driven.

Production systems may also be classified according to their main product(s) - meat, milk, eggs, manure and wool/fibre - as well as by services such as draught, transport and breeding. Systems may be further differentiated based on a scale ranging from "extensive" to "intensive", determined largely by their input-output relationships.

Livestock and their products fulfill different and often complementary functions. Apart from the marketed outputs, these include inputs to crop production (draught power and manure) and home consumption (subsistence), as well as their asset function and socio-cultural importance. At the farm level, therefore, livestock must be viewed in the context of the overall farm-household system, and the complex interactions involved must be understood technically as well as economically and socially.

Where livestock are an integral part of the farm household economy, flow products (draught, manure, milk), risk alleviation, capital accumulation and the utilisation of family labour with low opportunity costs are frequently the dominant production objectives. This is often in direct contrast to state objectives, which favour end-products (meat, hides, skins), and has been one reason why so many government livestock development initiatives have failed.

Strategy and Programme Formulation

A development strategy needs to anticipate and identify those forces that drive a particular livestock production system and the subsector as a whole. These can include population growth, market development (urbanisation and income growth), technological change and a changing resource base. An understanding of the direction and speed with which these factors may change, both at the sectoral and farm levels, is necessary so that future programmes can respond accordingly. As has been stressed, technological choice depends on the relative abundance of production factors, such as labour, land and capital.

Implicit in this process is the development of "what if" scenarios to anticipate the likely implications of any change. For demand-driven production systems, the variables usually include national and per caput incomes, population and population growth, product prices and price and income elasticities. For resource-driven production systems, trend projections that take account of anticipated changes in factor proportions, advances in technology and likely increases in productivity and resource availability are required. Any gap between supply and demand will have to be met either by expanding commercial production or by imports.

The first step is to develop the "without" intervention scenario by projecting future production and consumption trends. This will indicate gaps in supply, overuse of natural resources and import requirements, among other things. The next step is the formulation of development pathways (a combination of technical and institutional interventions) that address identified constraints, opportunities and any comparative advantages. Interventions then need to be evaluated under "with" and "without" scenarios, which involves sensitivity analysis of the critical variables as well as an assessment of social and cultural acceptability. Historically this has primarily been a top-down approach that has failed to take account of the aspirations and production objectives of producers, resulting in the poor performance of so many livestock development programmes. It is important, therefore, to ensure the active participation of all interested parties in the formulation process.

For animal production, possible programmes and policy interventions include the following broad classifications:

— On-farm interventions, adapted to specific agro-ecological conditions and production systems. These interventions commonly aim at increasing the availability or utilisation of local feeds, control of economically important diseases and/or improved housing and management.

— Institutional changes, including the structure and function of support services covering input supply, research, extension and training, processing and marketing and credit. Institutional programmes often complement technical interventions and aim at providing a support framework for livestock production that should be both cost-effective and congruent with overall government policies. The concept of "private" and "public" good will increasingly determine who will pay for such services.

— Genetic improvement programmes aimed at improving the livestock resource base. Options include within-breed selection of adapted indigenous breeds, substitution with exotic breeds or cross-breeding. The choice largely depends on the production system, its objectives and the resources at its disposal. Experience has shown serious misjudgement with policies aimed at importing exotic breeds with a corresponding neglect of indigenous breeds in many developing countries. Whatever breeding programme is adopted, equal attention needs to be given to the dissemination of improved genetic material. Dissemination concerns the institutional aspects as well as the choice of biotechnology, such as artificial insemination and embryo transfer.

— Animal health programmes aimed at limiting the impact of disease on animal production. Policy issues concern who will provide and pay for such services. Foremost is the need to control and protect the national livestock resource from major epizootic diseases, such as rinderpest, which is clearly a "public" good. Disease monitoring, veterinary investigation and legislation also fall within the public domain. On the other hand, control of diseases that cause production losses, for example, helminthiasis, is primarily a "private" good and determined on grounds of cost effectiveness.

— Processing and marketing policies related to investment in the necessary infrastructure that enables livestock products to safely supply existing demands as well as those of the future. Such specific issues as the design of structures, equipment, training and quality, along with the question of the degree of state intervention in the market, must also be addressed.

Impact Assessment

The predictive phase of the strategy design and planning process examines the various policy options and their implications in light of the government's broader development objectives. Strategy impact assessments usually address issues such as:

— economic efficiency;

— distribution and equity considerations (of both costs and benefits);

— stability (food supply, income, export earnings, etc.) and risk considerations;

— sustainability (environmental, financial and institutional);

— conformity with government objectives.

This is primarily a "bottom-up" procedure looking at the performance of the individual animal through to the production systems and finally the subsector as a whole. At the production system level, interventions can be assessed in terms of their economic, social and environmental impacts on households. It is at this level where interventions that are usually aimed at increasing productivity have to be shown to be both technically and financially viable. The results of these production system assessments may well necessitate changes in the off-farm elements of the proposed programme, such as marketing, processing, finance, extension and research.

Programme and policy implementation will also have implications at the sectoral and country levels. And some of these implications will have an impact on programme implementation as well, where they relate to the capacity of the ancillary services and industries to support and service the proposed development. These can be manipulated through price, monetary and trade policies, institutional reform and land regulations, which are mostly under direct government control. Those under private control can only be encouraged through supportive action. The impact of a successful outcome of the proposed programmes also needs to be assessed. Increased production will have an impact on income, consumption and labour, as well as consequences for the rest of the agricultural system.

As a result of the impact assessment, a set of development options can be presented to the government for consideration. Each option should be accompanied by a summary description, along with its likely impact and implications. It is the government that will make the final choices as to whether or not on certain strategies should be implemented and whether or not external assistance is necessary; ideally, the major donors would have participated in the review process.

2

Feed Resources for Livestock

Human society requires energy, food, a source of chemicals, shelter, and employment. There is good reason to believe that human food production can be sustained even with an increasing population. There are local food shortages but world supplies are more than adequate as evidenced by the fact that the major part of the world's cereal grain production is fed to livestock. By contrast, the present global level of energy use is not sustainable, as it depends on finite supplies of fossil fuel. Thus while, in the past, production of food for human consumption has been the priority activity, in the future, provision of energy will begin to take precedence. Present feed resources for livestock result mainly from activities directed towards human food production. In the future, they will increasingly be derived from the residues and by-products of energy production.

Fertile river deltas and valley floors will continue to be priority areas for rice production. However, pressures to increase yields and to reduce agro-chemical inputs will favour the introduction of rotations and associations with high biomass yielding crops that raise soil fertility while still contributing food and fibre. Sugar cane, with its unrivalled capacity for raising soil organic matter, the African oil palm and other multi-purpose trees are likely to become increasingly important components of these ecosystems as the search for renewable sources of energy increases in intensity.

Continuous rice culture as presently practised (with high-yielding varieties and agrochemical inputs) is not sustainable. This is confirmed by the results of a recent study from the International Rice Research Institute which shows that rice yield response to increasing fertiliser application is falling. Increasing the sustainability of rice production is most likely to be achieved by diversification. The traditional rice-duck-fish system has already proved to be an excellent biological control mechanism. Livestock manure was, and still is in many areas, the only fertiliser required if it is supported by diversified restorative cropping. But, as long as fossil fuel is cheap, animal traction will continue to be threatened by mechanisation, creating an incentive for increasing use of agro-chemicals. So begins the process of disintegration of the system, and dependence on outside inputs.

The Chinese deep pond and dyke system, sustained for thousands of years in the high water table region along the Pearl river, although demanding of capital for its establishment is probably the most productive and highly sustainable way of farming these otherwise difficult "water-logged" areas. Unfortunately, these systems which have sustained thousands of farmers over generations are also being threatened by the agro-chemical inputs of "modern" technology and the demands of an increasingly "consumer-oriented" society. On hillsides and sloping land at elevations below 2,000 m, sustainable land use requires the cultivation of perennial crops as the basic activity.

Systems of agroforestry are the only sustainable alternatives. They offer many options, including associations with short-cycle food crops and with grazing of livestock. But trees, in some form or other, must dominate such landscapes in the future. Finally, there are the arid and semi-arid regions where the only sustainable option is the cultivation of drought-resistant trees and shrubs, among which the *Prosopis* spp. have already shown their potential.

Categories of Livestolk Feeds

Four main categories of feeds will be dominant in tropical regions as pressure to carry out sustainable agricultural practices increases. These will be:

1. The fibrous residues from crops grown for human food production. Pre-eminent here will be the straws and stovers from cereal grains, mainly from rice but also from sorghum, millet and maize.
2. New feed resources derived from crops grown primarily as sources of renewable energy, or as contributors to soil fertility. In this category will be the range of products and by-products derived from sugar cane, African oil palm, the sugar palm tree, other multi-purpose trees (especially leguminous ones) and aquatic plants. These are rich in readily available energy (the juice from sugar cane and sugar palm, the oil and fruit from the African oil palm) or in protein (the leaves from multipurpose trees and aquatic plants).
3. Other by-products and residues.
4. Natural pastures, although their role is decreasing, will provide, for some time, the bulk of ruminant feed. However, there is a wealth of data on grasses and conventional forage crops. There will be little to be gained by doing more research with these feed resources.

A wide range of by-products and residues results from other food crops and cropping systems. In general, information exists on the nutritive value of by-products of oilseed and cereal milling. It is more important to know where the existing information can be found, which is the next point to be discussed.

Cereal Grains

Cereal grains are the staple food of the bulk of the world's population. The areas devoted to their cultivation, and the yields obtained, have steadily increased, mainly through the use of inputs derived from fossil fuel. Cultivation of cereals at these extremes of the yield response curve is not sustainable viewed in the context of the long term. However, cereal crops can be part of a sustainable cropping system, in which other crops are grown to restore the fertility exploited by the cereals. Ley (rotational) farming was introduced into Europe in the 18th century for this very purpose. The livestock farming network based around the rice crop, coordinated by IRRI is an example of this approach. Similar initiatives are needed in the other tropical continents.

In temperate regions of the world, cereal production exceeds what is needed for human consumption. The surpluses are fed to livestock and considerable amounts are exported. In contrast, most tropical countries import part of their needs of cereal grains for human consumption and many of them also import grain for feeding to livestock. In tropical countries, promoting the use of alternative feed resources, derived from crops which are more "environmentally friendly", will lead to more sustainable development and will increase self-reliance. The major objective of this manual is to encourage the use of these alternative feeds and, through research, to improve the efficiency of their utilisation at the level of small-scale farmers.

It should also be stressed that for many new "alternative" feed resources, learning how to grow them will have a higher priority initially than characterising their nutritive value; which is a way of emphasising that researchers concerned with developing sustainable livestock-based production systems in the tropics must be prepared to approach the problem in a multi-disciplinary fashion. For agronomic knowledge will often be an essential prelude to being able to use a particular plant as a livestock feed.

Cereal Grain Substitutes

Considerable advances have been made in the development of feeding systems using the products and by-products of sugar cane and the African oil palm, crops that with appropriate management have proved to be sustainable. The principal nutritional elements in these feed resources are sugar and oil respectively. There is insufficient knowledge about both these nutritional fuels, how best to complement them with protein and other support nutrients, and what their effect is on the composition and nutritional quality of the products when they are the basis of the diet.

Much work has been done with the cassava plant, designed to promote its use in livestock feeds. However, the major emphasis has been on producing a dry meal from the roots that could be used in conventional mixed feeds. This suited the interests of exporters and feed millers and compounders in the developed countries, but the added processing costs usually put the dry product outwith the financial means of small-scale farmers, who were usually the original growers of the crop. Future research with cassava and other root and tuber crops should be directed to simple ways of on-farm processing and conservation, so that the growers can also be the major beneficiaries.

A relatively new feed resource, especially appropriate for the dry and semi-arid ecosystems, is the fruit from trees such as *Prosopis spp.* Many of these tree fruits are rich in soluble sugars and gums. There is little documented information on their nutritive value or on ways in which they are being used or how they could be used better if managed in the most appropriate way. In general there is an urgent need to promote feeds and feeding systems that are farm-based, rather than relying on purchases of a "balanced" and "expensive" feed from a factory.

Alternative Sources of protiens

Oilseeds and pulses

The traditional sources of protein in the diets of monogastric animals are the by-products from oilseed milling and the processing of livestock, including fish. Although village-based processing is still the norm in some countries, the process of "development" has led to intensification and scaling-up of poultry and pig production in vertically integrated enterprises encompassing all aspects of the production cycle. This has made it difficult and costly for independent small-scale producers to acquire these feed resources.

There is an urgent necessity to develop protein sources that can be produced and processed on the farm. There is scope for the cultivation of traditional protein crops such as soya bean, groundnut, sunflower, as components of integrated and associated cropping systems. Unconventional legumes such as *Canavalia ensiformis* and *Canavalia gladiata* have received attention from researchers in Africa and especially in Latin America but, as yet, there is little impact at the level of the commercial farmer.

Multipurpose trees

Multi-purpose trees are essential features of sustainable development in all less-developed countries, but especially those in the tropics. Thanks to the enormous biological diversity within tropical trees and the accumulated experience of indigenous communities it is possible to advance rapidly in the identification, nutritional characterisation (including the presence of secondary plant compounds), cultural needs and growth patterns in associations with other species. There are hundreds of species in more than 40 botanical families and, within species, a wide range of provenances, many of which are already proving their "sustainability". To be able to take advantage of these "indigenous" riches is a stimulating challenge for researchers in the less-developed countries.

Research efforts should be concentrated on:

(i) identification of the secondary plant compounds in the leaves and how these are affected by soil, climatic and cultural practices;

(ii) developing ways of neutralising those compounds that have anti-nutritional effects (e.g., by simple ensiling and other fermentation methods);

(iii) increasing the availability to digestive enzymes of the amino acids in plant material consumed by monogastric species; and

(iv) finding combinations of leaves which maximise "by-pass" characteristics of the protein for ruminants.

Water plants

Where rainfall and/or irrigation are adequate, water plants are highly productive sources of protein-rich biomass and are ideal complements for fibre-free basal diets such as molasses, sugar cane juice and palm oil in pig and poultry feeding systems. As with the trees there are many species to choose from, each of which has its special characteristics which makes it more or less suitable for a given ecosystem. They satisfy the sustainability indicators and fulfill a particular niche because of their capacity to decontaminate water excessively charged with organic matter and plant nutrients.

The leaves of most water plants are more digestible than the leaves from trees and, generally, they appear to have low concentrations of anti-nutritional factors. The problem in practice has proved to be more in the area of agronomy than in nutrition. It is almost certain, that at the present time, learning how to grow water plants in continuous culture has a higher priority than characterising their nutritive value.

Unicellular protein

It is time to return to this subject but from the point of view of natural ecosystems rather than the fossil-fuel based schemes which have largely failed because of health-related issues (e.g., yeasts using petroleum derivatives as substrate) or cost (e.g., torula yeast from sugar cane derivatives). *Spirulina* has the potential to reproduce at a high rate on biodigester effluents, and has an amino acid pattern ideally suited to complement low-protein, energy-rich feeds derived from tropical resources such as sugar cane and African oil palm. Recent developments on the growing of *Chlorella* and *Snedecus* algae show promise as the basis for a low-cost method of harnessing solar energy and atmospheric nitrogen fixation to produce high-quality protein; potential rates of productivity of 9,000 kg protein/ha/year were reported by Chowdhury *et al.*

Worms, insects and larvae

Cultivation of the Californian Red Worm on livestock excreta has proved to be a commercially viable method for producing concentrated organic fertiliser (humus) on farms in Colombia. The worms, which are really a by-product of this process, are a source of high quality animal protein which has potential as a supplement for poultry, especially at family farm level. This same medium (livestock excreta) can be used for cultivation of a range of insects and larvae. Peasant farmers on the Pacific coast of Colombia are skilled in disseminating termite mounds in the forest and harvesting the contents as feed for poultry. There is much to be researched in this category of unconventional protein-rich resources.

Silage from animal and fish by-products

In industrial countries there are well developed technologies for recovering by-products of animal and fish processing and converting them into protein-rich meals. However, such facilities are rarely found in tropical less-developed countries, especially at the level of small towns and villages where slaughtering and processing of fish are done in rudimentary conditions and where by-products often become contaminating wastes. In such situations, the ensiling of the by-products, using molasses and crude syrups derived from sugar cane, is a simple and appropriate method of conservation.

Small non-ruminant herbivores

Cardozo lists the following species as priority candidates for research and development: rabbits (*Orictulagus cuniculus*), guinea pigs (*Cavia porcellus*), geese (*Anas anser*), iguanas (*Iguana iguana*), picure (*Dascyprocta spp.*), capybara (*Hydrochaerus capybara*) and snails (*Pomacea spp.*).

The advantage of the majority of species in this category is their high reproductive rate and their capacity to select and utilise plants which would not normally be fed to domesticated livestock. Rabbits, snails and guinea pigs, are traditionally managed in confinement. In fact, such a practice is essential in the case of rabbits and snails which rapidly become serious pests of food crops if allowed to proliferate without control. Geese are efficient grazers and their management should probably be based on some form of scavenging system. For the other species, domestication is almost certainly undesirable, and their exploitation in natural habitats is the preferred approach.

Multi-purpose cattle and buffaloes

The concept of multi-purpose use of large ruminants is gradually gaining acceptance, not least because of the technical and socio-economic failures of most projects that aimed to exploit specialised breeds and management systems. Two main focal points can be identified: the mainly dual-purpose management of crossbred *Bos indicus* x *Bos taurus* breeds for milk and beef production and the triple-purpose management of buffaloes and mainly *Bos indicus* breeds for work, milk and beef. Much more research needs to be done with these breeds and production systems.

From the animal nutrition standpoint, interesting areas of work relate to selection, offer level and quality of the predominantly fibrous feeds, the efficient use of which is the prerogative of these breeds and systems. The comparative advantage of the buffalo is still not well understood and even less well documented. Equally the advantages of restricted suckling of the calves, although highly appreciated and well understood by tropical farmers, poses questions to researchers concerning interactions between animal behaviour, productivity and feed utilisation efficiency, as this affects both dam and offspring.

The effect of work on productive traits of cows is highly correlated with nutrition. However, much of the research in this area has been done with the aim of characterising nutrient requirements rather than understanding how best to manipulate the system in order to optimise use of local feed resources and 'catalytic" supplements.

Nutrition of Non-ruminants

Protein Supplementation

With the exception of the organic waste of urban origin, all the feed resources mentioned above are characterised by very low levels of protein in the dry matter ranging from 1% in sugar cane juice and molasses to 5% in whole African palm fruit. One consequence of this low level of protein in the basal diet is that almost all the requirements for amino acids have to be supplied in a supplement. While this could be considered to be a disadvantage, in fact the contrary may be the case because:

— It is much easier, and may be cheaper, to find a supplement already balanced with respect to the essential amino acids (lysine, methionine, etc.) than to make a supplement that will compensate for the imbalanced amino acid profile present in cereal grains.

— The total amount of protein to be offered is less because when it is derived almost wholly from supplements, rather than partially from cereal grains, it is better balanced in the essential amino acids.

These possibilities can be visualised more easily in the comparisons presented in Figure 1 which show that the essential amino acid requirements of sows can be contained in 30–35% less total protein, than what is recommended by NRC which is inflated by the presence of excessive amounts of the non-essential amino acids present in cereal grains.

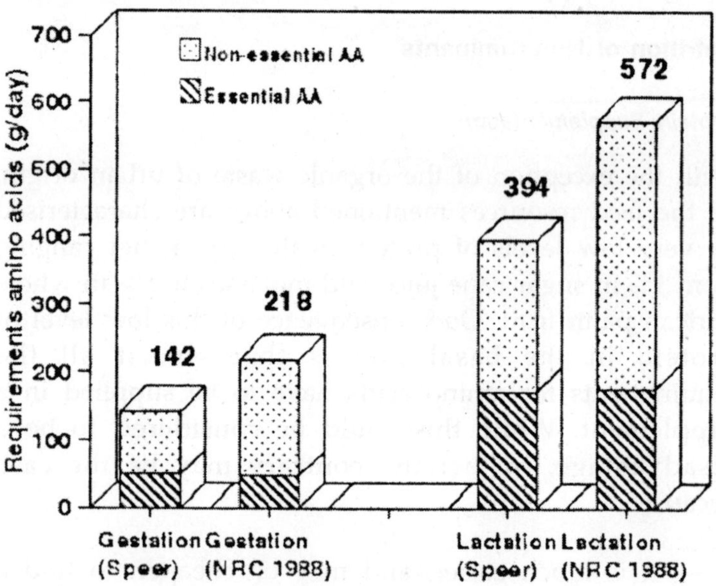

Figure 1. Protein requirements for pregnant and lactating pigs.

The final point is the relationship between the amount of the supplement and the level of animal performance. This is especially important in most tropical countries where the ratio - price of protein : price of energy - is much wider than is the case in temperate countries. Thus the point on the response curve where profit per fattening pig is optimised will not necessarily coincide with the point where biological performance is maximised. An example taken from Vietnam clearly illustrates this point (Figure 2).

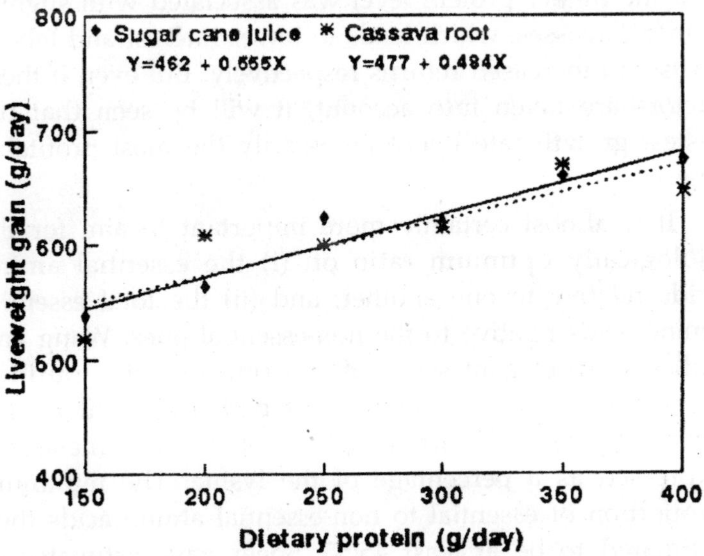

Figure 2. Effect of dietary protein supply on growth rate of pigs fed basal diets of sugar cane juice or cassava root meal.

The regression equations show that for every 100 g/day increase in supply of protein liveweight gain was increased by 55.5 g/day on sugar cane juice and by 48.4 g/day on cassava root meal. Putting the value of liveweight at US$1.00/kg and the cost of soya bean at US$400/tonne then a liveweight gain of 55.5 g/day, worth US$0.055, requires 200 g soya bean meal which costs US$0.80. Clearly there is no economic advantage in giving more than the minimum amount of soya bean meal which will support a liveweight gain of 542 g/day. This is well below the optimum biological response of 684 g/day liveweight gain obtained with 800 g/day of soya bean meal.

The calculation is a little more complex as faster growth means less time for reaching slaughter weight, and the higher protein level was associated with slightly leaner carcasses, which leads to lower interest and labour costs and increased returns respectively. But even if these factors are taken into account, it will be seen that the fastest growth rate is not necessarily the most profitable one.

It is almost certainly more important to aim for the biologically optimum ratio of: (i) the essential amino acids relative to one another; and (ii) the total essential amino acids relative to the non-essential ones. Wang and Fuller in an elegant series of experiments using purified diets have provided estimates of both these criteria for growing pigs. In Figure 3 the estimated requirements are expressed as a percentage of the lysine. The minimum proportion of essential to non-essential amino acids they estimated to be at least 45:55. Speer gave estimates of 35:65 and 44:56 for gestating and lactating pigs respectively.

The recommended strategy when balancing tropical energy-rich feeds, is to prepare a protein supplement which has a balance of amino acids that resembles as closely as possible that proposed by Wang and Fuller (Table 1). Where a variety of protein sources are on offer, a "least cost" computer program can be used to achieve such a balance. An example of this approach was given by Speedy et al. for preparing a protein supplement to balance sugar cane juice for pig fattening in Swaziland.

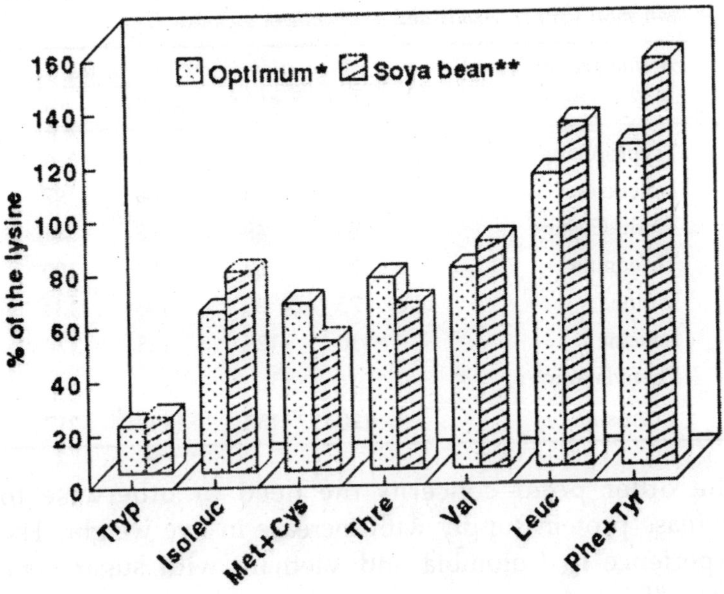

Figure 3: Amino acids in soya bean meal: comparison with ideal protein

For the next step one should ideally be able to refer to regression equations relating productive function to protein for the particular energy-rich "tropical" feed being used. An example of this approach is given in the work reported by Bui Huy Nhu Phuc *et al.* and Ospina *et al.*, referred to above (Figure 2). Response curves relating growth and carcass traits with supply of a balanced protein (soya bean meal), were determined for basal diets of sugar cane juice and cassava root meal, respectively. From such data it is possible to calculate the marginal advantage (or disadvantage) of a particular level of protein input in terms of the expected increase in productivity that will result, compared with using a lower level of protein.

Table 1. Balance of essential amino acids in "ideal" protein and that from soya bean meal (SBM), Azolla and Trichanthera gigantea (Tg)

Amino acid of lysine)	Proportion in the protein (as %			
	"Ideal"	SBM	Azolla	Tg
Tryptophane	18	21	31	NA
Isoleucine	60	75	60	97
Meth+Cystine	63	49	63	81
Threonine	72	62	72	100
Valine	75	85	103	134
Leucine	110	129	138	172
Phenylalanine +Tyrosine	120	152	149	217

The other point concerns the need or otherwise to increase protein supply with increase in live weight. The experience in Colombia and Vietnam with sugar cane juice, "B" molasses, cassava root by-product and oil palm derivatives, as the basal diet, is that it is simpler for the farmer to give a fixed quantity of protein daily and that there is no economic advantage in having varying levels according to liveweight. The allowances presently used are set out in Table 2.

Table 2: Allowances of "ideal" protein for pigs in different phases of the production cycle

Category protein	Liveweight (kg)	"Ideal" allowance (g/day)
Growing-finishing	20–90	200
Gestation	>90	150
Lactation	>90	350

These quantities are based on the assumption that the essential amino acids will supply at least 45% of the total amino acids and that the essential amino acids will be balanced according to the recommendations of Wang and Fuller (Table 1). The fact that the regression coefficients linking protein supply with liveweight gain were almost the same for cassava root meal and sugar cane juice (Figure 1) indicates that probably other low-protein, low-fibre energy feeds will respond in a similar manner. Support for this hypothesis is found in the results with oil-press fibre fed as the basal diet to fattening pigs. In an experiment with levels of from 500 to 1,000 g/day of a fortified soya bean supplement, there was no biological advantage from giving more than 200 protein g/day throughout the growing-fattening (20–90 kg) period.

It is interesting, in the light of the above recommendations, to refer to an experiment carried out over 30 years ago (1957/58) in Louisiana State University by Thrasher *et al.* These researchers offered growingfattening pigs, balanced with soya bean meal, meat meal and alfalfa meal to provide 16 declining to 14% protein; a 6th group of pigs had free access to raw sugar, maize meal and a protein supplement.

Growth and feed conversion on all treatments were similar but the pigs on the free choice system only consumed 258 g protein/day, of which 95 g came from the maize and only 163 g from the protein supplement. These same pigs consumed 45% of their diet as sugar. The pigs on the mixed feeds consuming this same level of sugar consumed 400 g protein daily yet they performed no better than the free choice group which consumed 36% less protein. It is unlikely that, on all occasions, it will be possible to prepare an "ideal" protein supplement with the perfect complement of amino acids. For all practical purposes, the protein of soya bean is sufficiently close to the ideal protein (Figure 3) as to be the supplement of choice where it is available or can be grown.

In fact, on small farms in remote areas, the availability of protein sources may well be restricted to what can be grown on the farm or at best complemented with some by-product produced in the nearby village. In these situations it is important to have some idea of the amino acid balance of feeds that can be grown locally. It is enough at this stage to point to the examples in Figures 4 and 5 of two foliar sources of protein which can be grown in the tropics: the water fern *Azolla* spp. and the multi-purpose tree *Trichanthera gigantea. Azolla* is readily consumed by pigs, especially during pregnancy. The data in Figure 4 indicate that the protein in *Azolla* has an excellent balance of amino acids, better even than soya bean. In this case the limitation to its use will be voluntary intake and the digestibility of the protein.

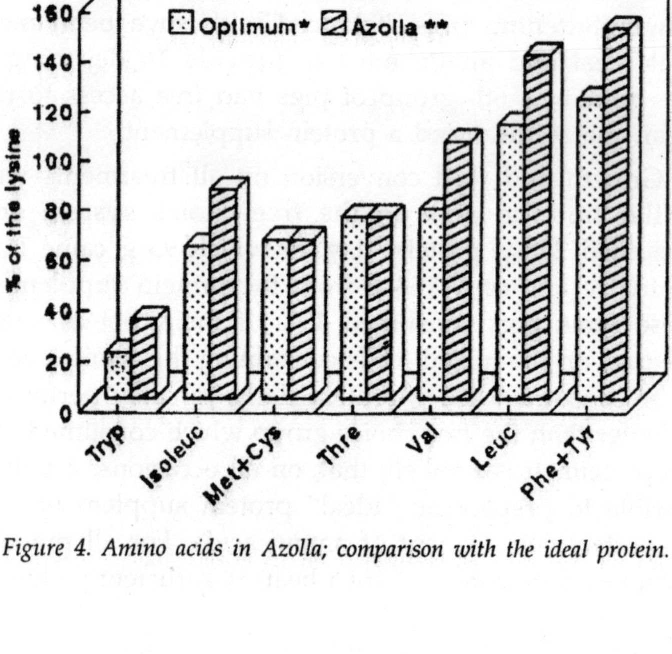

Figure 4. Amino acids in Azolla; comparison with the ideal protein.

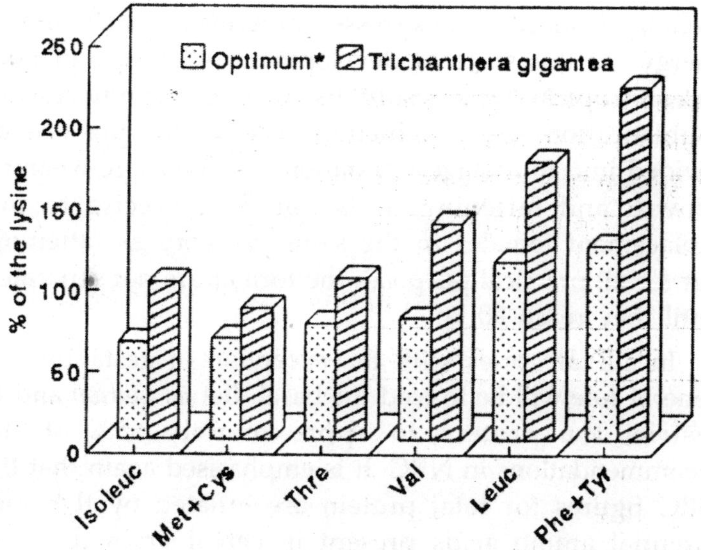

Figure 5. The essential amino acids in Trichanthera gigantea as percentage of lysine, compared with the ideal protein of Wang and Fuller.

The same is true, although to a lesser extent, with *Trichanthera gigantea*. The protein in the leaves has a good amino acid balance and according to some early work there are few secondary plant compounds. Pigs eat it well, especially during pregnancy. However, even when eaten in amounts that theoretically supply all the protein needs, pregnant pigs rapidly lost body condition when given only *Trichanthera* as a supplement to sugar cane juice. Up to 30% replacement of the soya bean protein by *Trichanthera* appears to be feasible, but either low digestibility or presence of secondary plant compounds appeared to limit its use at higher levels.

Energy Allowances

The energy allowances for pigs in the classical tables of feeding standards are expressed in terms of metabolisable energy. At this stage of development of the use of non-cereal tropical feed resources, this is an unnecessary sophistication. The production cycle of the pig can be divided into four stages: gestation, lactation, pre-weaning growth and fattening. It is not very precise to put replacement females in the same category as fattening, but for all practical purposes the former are not separated until they reach 90 kg.

In all categories with the exception of gestation, the general rule will be to feed the basal diet *ad libitum* and to restrict the protein to approximately 65% of the recommendations in NRC. It is emphasised again that the NRC figures for total protein are inflated by the non-essential amino acids present in cereal grain diets on which NRC standards are based. When low-protein tropical feed resources are used, the ratio of essential to non-essential amino acids can be set at close to the optimum (45:55) because:

— the total amount of amino acids in the basal diet is low therefore most of the amino acids comes from the protein supplement which is usually well balanced, and

— the protein that is present in tropical energy-rich feeds tends to have a better amino acid balance than the protein in cereal grains.

In gestation the feed allowance will be determined by the energy density of the basal feed. The aim should be to use feeds with low nutritional density so the animal can have free access to at least one component of the diet. In this way, the "hunger syndrome" is avoided. The result is a "contented" sow and the support, rather than the opposition, of animal welfare activists.

Minerals and Vitamins

In this case the indicator of sustainability is "minimum" dependence on outside inputs, especially those sold by feed mills and other intermediaries. Careful selection of the supplement can help to eliminate part of the need for both minerals and vitamins. For example if some palatable and digestible green foliage can be given free choice the pigs will obtain from it most of the vitamins except for D which is abundantly provided through the and trace minerals they need. The major minerals provided by the commonly used protein supplements are shown in Table 3.

Table 3: Amounts of major minerals supplied by the protein supplement as soya bean meal alone, or in combination with fish meal, Azolla or Trichanthera gigantea.

Protein source (g/d)	*DM (g/d)*	*P (g/d)*	*Ca (g/d)*	*K (g/d)*
Soya bean meal 50:50 SBM:	450	1.3	1.3	6.7
fish meal	425	7.3	13	5
Azolla:				
30% fattening	200	1.4	3.4	3.6
100% pregnancy	500	3.5	8.5	9.0
Trichanthera gigantea				
30% in pregnancy	200	0.8	4.8	7.6
Requirements (g/d)				
Fattening	9–12	11–16	2–3	
Pregnancy		11	14	4

When a mixture of soya bean meal and fish meal (75:25) is used as the protein source it will provide much of the needed calcium, phosphorus and sodium. Sources of calcium (lime) and sodium (salt) are available in most farm households and rural villages. In some cases the phosphorus can be supplied as phosphate fertiliser. Useful data on vitamin and mineral content of tropical feed resources can be found in Tropical Feeds.

Non-Cereal Grain Feed Resources for Poultry

As with pigs, it must be made clear at the outset, that the purpose of this section is to assist those researchers who believe that poultry and water fowl production should be carried out in a sustainable way. Feeding systems based on cereal grain will not therefore be discussed. In any event, there is a wealth of information on such systems in the classical text books and bulletins in developed countries.

There appear to have been no attempts to re-examine the nutrient needs of poultry in the light of alternative non-cereal grain feeding. One promising development, instigated in the University of New England, Australia, is based on the premise that, given the opportunity to select, poultry are able to balance their diet according to productive need and environmental stress. Cumming and his colleagues have reported economies in feed and no loss of production with this free-choice management system.

The disadvantage of this approach,in the context of tropical feed resources, is that protein-rich feeds in such regions are generally much more expensive than energy-rich feeds. Thus it is almost never profitable to feed protein at a level which will maximise animal performance. For this reason protein almost always will be rationed to secure the most profitable response.

There has so far been no commercial development of poultry feeding systems which use non-cereal tropical feed resources. Work with ducks is the most advanced in this regard. But the principles are the same. If tropical feed resources of low protein content are used then the protein supply, balanced according to the "ideal" protein, can probably be reduced to two-thirds of the recommended allowances which, as for pigs, were worked out against the background of cereal grain diets. A factor which may have to be taken into account, especially with ducks, is the need for sulphur-containing amino acids for feather growth. The testing of this hypothesis is recommended as a high priority research area. The other principles set out for pig feeding on tropical feed resources, with regard to energy, minerals and vitamins, apply equally well to poultry including water fowl.

Nutrition of Non-ruminant Herbivores

The species which are raised commercially in a number of tropical countries are rabbits and guinea pigs. Unfortunately, as in the case of poultry, the intensive "modern" feeding systems are invariably based on conventional, mostly "non-tropical" feed resources and no serious attempt has been made to re-assess their nutritive requirements when alternative "tropical" feeds are used. Advances are now being made in the design of feeding systems for rabbits, using tropical feeds, but as yet there are no reports of possible savings in protein requirements, for example that might be achieved y replacing cereal grain with low-protein sugar or oil-rich energy feeds. Theoretically, savings should be possible as the same arguments concerning digestive processes apply to non-ruminant herbivores as to pigs. There is an urgent need to make such studies in the newly developed feeding systems being used for rabbits, and to begin comparable research with guinea pigs.

The most recent version of the recommendations of the ARC on nutrient requirements of rabbits and guinea pigs was published in 1977. Since this date, much research has been done on the digestive physiology and metabolism in these species, although it has not been interpreted in terms of recommendations on nutrient requirements. It must also be made clear, however, that almost always this research was done using conventional "temperate" feed resources, which are neither available nor sustainable in the way they are produced in temperate countries.

On such feeding systems the objectives of maximising biological efficiency led progressively to a reduced capacity to utilise herbaccous plants. Emphasis was put on optimising availability of glucose and amino acids, which was favoured by the use of diets with high levels of starch and preformed protein. The role of microbial synthesis in the caccum and the utilisation of this source of amino acids by coprophagy was largely ignored.

For these reasons, researchers in tropical countries are encouraged to take advantage of the opportunities offered in the area of nutrition of "small non-ruminant herbivores" in the tropics, and to participate in the development of feed resources and feeding systems which take account of sustainability criteria. The appearance of a new computerised journal dealing specifically with small non-ruminant herbivores reflects the increasing interest being given to these species.

The nutritional requirements of small non-ruminant herbivores are determined by the relative activities of:

— digestion and absorption in the duodenum, and

— digestion by fermentation and absorption in the caecum.

— In the case of the rabbit, the guinea pig and probably the picure, must be added coprophagy as a physiological recycling mechanism.

The last two processes are ruled out in the case of the snails at least until more is known about the physiology of digestion in these species. The fact that the small non-ruminant herbivores must satisfy their nutritional requirements by pathways (i) and (ii) has important implications, for example: eergy metabolism is based on the utilisation of glucose, VFA and lipids. Therefore, tropical feed resources must be selected according to their capacity to supply: digestible fibre, sugars and starch and pre-formed lipids. In the tropics, satisfying the energy needs of small non-ruminant herbivores in a sustainable manner probably requires that major emphasis is given to maximising intake rather than digestibility.

3

Issues in Landless Livestock Ruminant Production Systems

Landless Livestock Ruminant Production Systems (LLR) are a solely livestock systems (less than 10 % income from non-livestock agriculture), where less than 10 % of the feed dry matter fed to the animals is farm produced, the annual average stocking rates are above 10 LU per ha of agriculture land, and the value of the ruminant enterprises is higher than that of the pig or poultry enterprises. In LLR systems, the feed is mainly introduced from outside the farm system, thus separating decisions on feed use from those on feed production, and particularly decisions on the use of manure on fields to produce feed and cash crops. These systems are very open systems in terms of nutrient flow; the nutrients cannot be circulated within the farm system.

Figure 1 presents the inputs and outputs of LLR systems. They share this feature with the landless monogastric livestock systems (LLM), the main difference is that ruminants need a fibrous ration whereas in monogastrics the feed conversion of concentrates to live weight gains is substantially more efficient. Thus LLM systems are only competitive in situations where cheap concentrates are available and consumers are prepared to pay substantially more for quality beef than for chicken or pork.

Inputs	*Outputs*
Feed Resources	*Products*
(roughage, concentrates, byproducts)	- meat
- N, P and other minerals	- hides
- additives/vitamins	- slaughter wastes
- heavy metals- manure	
Primary Production LLR Systems	
Animal Resources	*Emissions*
Animals of various age and weight from land-based systems	To the soil: - N, P, other minerals
Drugs	to surface water: - OM, N, P, other minerals
Fossil Energy	to the atmosphere: - local importance: NH_3 - global importance: CO_2, NO_x, CH_4

Figure 1. Scheme of inputs and outputs in LLR systems.

LLR systems are concentrated in a few regions of the World. Cattle LLR systems are found in a few OECD countries, Eastern Europe, CIS and near the main cities in developingcountries, while sheep LLR systems are foùnd in West Asia and North Africa (WANA).

Livestock resources: LLR systems are part of a stratified livestock production system. Livestock in LLR systems originate from other (land-based) systems, e.g. offspring from range fed. beef cattle are finished in intensive feedlots; lambs from the pastoral areas are fattened intensively; male calves from dairy breeds are fattened on milk; and high yielding dairy cows from the rural areas are sold for a final lactation in urban dairies. LLR systems are largely interlinked with other livestock and mixed production systems.

Feed resources: Feeding is mainly based on good quality high energy feeds with a maximum intake of concentrates and a minimum intake of roughage. Although grains and protein supplements are the main sources of concentrate feed, depending on the system and the region, by-products from the agro-processing industry could be important. Minimum roughage requirements are around 15-30% of the ration.

LLR Subsystems

The following subsystems and their geographical distribution are recognised within LLR systems:

(1) *Feedlot fattening and finishing of beef cattle:* This system is mainly concentrated in the USA and particularly in the states of Texas, Nebraska, Kansas, Colorado and California;

(2) *Veal production:* This type of production is mainly found in the EU and concentrated in the Netherlands, France and Italy;

(3) Intensive finishing of sheep, mainly in the Middle East, e.g. in Syria and Iran;

(4) Large-scale beef (and dairy) production in EE and CIS;

(5) Urban milk production in developing countries, e.g. cities on Indian subcontinent.

Beef Fattening in Feedlots

Feedlot fattening of beef cattle is mainly done in the USA, though this practice of upgrading beef quality also exists in the EU. Furthermore, in few Latin American and

African countries some finishing of cattle in feedlots is apparent, mainly to benefit from compensatory growth effects. However, in these last-mentioned regions the fattening of beef cattle is nearly always an economic activity next to arable farming or land-based livestock production.

In the USA, the beef industry is an important segment of livestock production. The increasing population and the rising consumer buying power have together contributed to an increase in demand resulting in relatively favourable prices for beef. Consumption of beef per capita increased from 29 kg in 1946 to 50 - 56 kg in 1975 - 1980 and stabilised at around 50 kg in 1990. The practice of grain feeding of cattle has increased rapidly and over 90 % of the steers and heifers slaughtered are marketed by feedlots. Primarily as a result of the rapid increase in the demand for fed beef and increased financing from sources outside agriculture, fewer but larger feedlots have evolved with a marked geographic concentration.

Beef production is very sensitive to changes in profitability induced by changes in beef prices and consumption, cattle prices and feed cost. As most steers and heifers are slaughtered at least 3 years after their conception, a cyclical pattern of beef production results.

Generally two types of feedlot enterprises can be recognised:

(1) *Farmer-operated feedlots:* generally with a one-time capacity of less than 1000 head of cattle. Typically the feedlot farmers with under 1000 cattle are also involved in other farming enterprises, especially pig production and beef cows. Feedlots with a capacity of less than 200 head are often only in operation during the non-cropping season when

off-season labour is available. This type of farmer-operated feedlots can be found in almost all cattle feeding regions of the USA especially in the central part of the country.

(2) *Large-scale feedlots:* feedlots with a capacity of more than 1000 head which are in operation all year. The importance of this type of feedlot is ever increasing; in 1980 ca. 75% of the marketed feed cattle passed through this type of feedlot, compared to ca. 84% in 1990. Ownership ranges from sole proprietorships to corporate farms, including cooperatives. Most feedlots of this category purchase all or most of their feeder cattle, feed and other inputs. Feedstuffs are premixed and delivered to the cattle feed troughs by self-unloading trucks or other power equipment.

The large-scale feedlots all come under the landless category. Of the farmer-operated category of feedlots no accurate data on land - livestock ratio are available. However, according to the definition of this study, farmers with farmland for beef cows and probably arable land for feed production cannot be classified as landless and will therefore, not be discussed any further. Noteworthy, is that the high animal concentration on these farms could still form a potential environmental problem. Therefore all cattle fattened in feedlots are considered as landless, though strictly speaking this is probably an overestimation of the size of LLR systems.

Sources of cattle: Feeder cattle are produced in almost all regions of the USA and there is a considerable inter-regional movement of feeder cattle from the place of origin to the place where they will be fattened. Some rather definite flow patterns of feeder cattle have been established, but many travel a rather circuitous route from birth to final destination.

Around two-thirds of the cattle fattened are steers, the remainder being heifers and less than 1 % are cows. This ratio changes depending on 'the beef cycle'; a situation of an expanding or a shrinking beef breeding herd. Approximately 75 % of the cattle are English breeds or crossbreeds; around 10 % are dairy breeds and crosses. The initial weight depends, among other things, on the price of feed grain; low prices make it attractive to start with lighter weight cattle for a longer grain feeding period.

Basically 2 systems of beef operations can be recognised:

(1) *Cow calf operation:* The calf is placed in the feedlot when about 180 kg and less than 200 days old. Feeder is fed to 475 kg during 280 days.

(2) *Cow calf range operation:* The calf is weaned and put out on range at an approximate weight of 180 kg; age less than 200 days. Feeder (or stocker) is left on range till it reaches a weight of 385 kg at around 600 days. Feeder is finished in feedlot to 475 kg in around 120 days (720 days of age).

Feeding: The composition of the rations fed in the finishing operations depends largely on the types of feed produced local availability and on weights and grades of feeder cattle. Rations can range from high-roughage low energy rations to high-energy rations composed almost entirely of concentrates. Farm-operated feedlots tend to feed a higher proportion of roughage than do larger lots. Examples of ingredients in rations are:

— maize and maize silage with soya bean meal and urea (Corn Belt and Lake States);

— barley, maize silage, by-products feed (e.g. apple pomace, potato waste, sugarbeet by-products); and

— maize, sorghum grain, alfalfa, straw, cotton seed hulls and molasses.

By-products are chiefly fed in the large feedlots. Some hormone-like growth stimulators, antibiotic feed additives and ionophores (rumen altering factors) are legalised to be included in the feed and are very commonly used. Average production parameters and ranges are given in Table 1.

Table 1. Average production parameters

Production parameter Reasonable	Range	average
Starting weight (kg)	260 - 400	340
daily gain (kg)	0.75 - 1.3	0.9
feed conversion	6 - 11	8
fattening period (days)	90 - 280	140
final weight (kg)	410 - 550	475
vacant days (d/place/year)	10 - 100	

Housing: Beef cattle in feedlots do not require shelter for protection from cold weather. However, in the Southern part of the USA more efficient growth rates will be achieved if shade is provided. There is a general trend to keep the cattle in semi-confinement in order to control waste production. Where cattle are confined and housed over slatted floors, all faeces and urine can be collected, thus eliminating the need for using bedding material. At present most of the cattle in feedlots are still kept on concrete floors, or in dry regions, on an unpaved area. Solids from manure are either collected daily and stored, or allowed to dry in the feedlot and removed periodically before spreading on fields. Collection of urine is limited to feedlots with a slatted floor.

Veal Production

Veal production is mainly prevalent in the EU, where it was highly stimulated by subsidised milk powder used for calf rearing. Veal calves are exclusively given milk or milk substitutes in order to keep them at the pre-ruminant stage, thus avoiding the development of the forestomachs. Slaughtered at 2 - 5 months of age, at a live weight varying from 100 - 250 kg, they must grow rapidly (over 1 kg per day) and provide a high dressing percentage a well-conformed and sufficiently fat-covered carcass with pale pinkish meat. This last-mentioned characteristic is very important from a commercial point of view (high premiums are payed by consumers for this type of meat) and can only be obtained from pre-ruminant animals on an iron deficient diet. Two main types of veal production can be distinguished:

(1) *Nursed veal calves:* This system is limited to two areas in France and include less than 12 % of veal calves slaughtered in France. The calves are mainly offspring of French beef breeds and crosses with dairy cattle. Almost all calves are born on the farm where they are reared, but sometimes a few additional calves are bought. The price of this top quality product is high. This system is decreasing because it gives a lower income than suckling calves at pasture which are sold at weaning for further fattening and red meat production.

(2) *Veal calves reared on milk substitutes:* The production of milk substitute-fed veal calves is mainly localised in Western and Southwestern France, Northern Italy and Central Netherlands. Friesians with an

increasing proportion of Holstein blood predominate. Most of the calves are born outside the farms where they are reared. They are usually bought from various sources at 1-3 weeks of age, Italy imports calves for rearing from The Netherlands and France. A significant number is imported into mainland Europe from the UK. Over 85 % of the production is organised under contract by dairy and non-dairy companies manufacturing milk substitutes as well as slaughterhouse companies.

The first system is a land-based system, where veal production is only a minor activity. The second system is primarily a landless system, where veal production is the main or sole agriculture activity. Only the landless system will be discussed in this study.

Sources of calves. Male and surplus female calves are from dairy breeds, and calves from dairy cows inseminated with semen from beef breeds.

Feeding and growth. Milk substitutes fed to veal calves mainly contains milk ingredients (milk and whey powder; 70-90% of total). Starch derivates, fat substitutes (e.g. tallow, lard and saturated vegetable oils), minerals and vitamins are added in various proportions. Growth rates of an average 1.2 kg per day are common (0.7 - 0.8 kg during the first month increasing to 1.4 - 1.6 during the last part of the fattening period) with an average daily consumption of 1.8 - 2.0 kg milk substitute powder per day.

The milk substitute ration is highly digestible and the transformation in edible products is very efficient, however, the milk substitute is an expensive feed. The development of the production of milk substitutes has resulted from the growing surpluses of skim milk and whey powder in the EU Milk substitutes play an important role in the regularisation of the dairy market.

The use of anabolic agents to improve nitrogen retention, live weight gain (LWG) and feed conversion rate (FCR) are strictly not permitted in the EU but in reality are used frequently.

Average production performance of veal calves in the main EU countries is in Table 2.

Table 2. Production performance of veal calves in the EU in 19872 .
Production performance of veal calves in the EU in 1987

Country	*Fattening period (days)*	*LWG (kg/day)*	*FCR (kg DM/kg)*	*Carcass weight (kg*
France	120	1.2	1.45	113
Netherlands	180	1.1	1.60	144
Italy	150	1.2	1.50	130

Housing. Veal calves are generally housed individually in boxes on a slatted floor. There is increasing public pressure for animal welfare reasons to house veal calves in groups. The manure from veal calves is a liquid slurry with a high water content (about 98%).

Sheep Fattening

Sheep production in WANA is mainly based on the traditional pastoral system making use of the vast desert ranges and natural pastures. Following migratory or semi-sedentary grazing practices, the pastoralists and their livestock trek through the harsh terrain to produce milk and meat from resources that would not have been used otherwise. Milk production is important in traditional pastoral sheep production. With the rising demand for sheep and goat meat the pressure on the rangeland has increased with increased risks of overgrazing. Purchase of lambs for fattening on the available feedstuffs on a farm has been a common practice in the region.

Large-scale feedlot enterprises based on purchased sheep and purchased feed have been stimulated by rising meat prices and/or by government development programmes. In Iran, for example, intensive sheep fattening has increased because of:

(1) the low productivity of the land-based system; and

(2) the provision of subsidised barley to increase the level of self-sufficiency in mutton.

Prevention of further degradation of rangelands is often the justifiable objective of the barley subsidy.

Galal reports that in Egypt fattening operations become particularly active 2 to 4 months before Eid Al-Adha. Rams weighing 20 to 40 kg are fattened to over 50 kg. In each feedlot, sheep numbers range from as few as 2 to over 100. Sheep are sold at local markets, to butchers or to urban centres for religious festivals.

There are clear peaks in demand for mutton during the muslim religious festivals, and as the dates of the festivals change over the years, so do the fattening periods.

Sources of animals. Male lambs, surplus female lambs but also older (cull) animals are purchased from pastoralist herds and small farmers for further intensive fattening over a period of 2 to 4 months.

Feeding and growth. Feeding is based on feed grain (often subsidised), supplemented with by-products and cut forage or straw. The proportion of roughage although often not well described, is known to be generally very low (10-30%). Little data is available on the performance of sheep in feedlots.

Housing. Little information is available on housing, but it is assumed to be a very simple system: mainly a fenced off, unpaved area with some shelter to protect the sheep from sunshine. There are no indications in the literature that collection, storage and disposal of manure from feedlots is causing major problems. Manure is generally considered to be a valuable product and sold for use in crop production.

Large-scale Beef Production

Large-scale beef production used to be entirely in the hands of the state and cooperative sector. In Eastern Europe the average size of state farms varies between 500 and 7000 ha and in the cooperative sector between 100 and 4500 ha. Herd size varied from around 200 head in former Yugoslavia to over 1000 head in Rumania. In the CIS, the state and collective sector with over 52,000 farms owned almost 80 % of the cattle population.

The average collective farm size in the Russian Federation is around 6,600 ha and 1,900 head of cattle. State farms are around 9,000 ha (50 % planted) and around 2,100 head of cattle. More than 90 % of state and collective farms raise cattle; most dairy cattle are not grazed or only grazed during the short growing season. Little use is made of existing topographical potential for the stratification of the production system (e.g. raising young dairy stock in the hills and mountains). Fattening is mostly done on the same farm where young stock is also reared.

Feeding efficiency is low; livestock productivity is only about 50 - 60 % of Western European levels. This is partly due to low quality of the roughage, imbalanced feeding by over-using grain, and cereal by-products (ca. 40-50% of the ration), resulting in protein shortages. In the CIS for example, only 13 % of the feed requirements are met from pastures and a further 21 % from roughage (hay, silage and straw). Considering the region with such vast areas suitable for pasture, this is an extremely low percentage.

On the other hand, inefficiencies of livestock production in EE and the CIS are sometimes exaggerated. Mudahar et al. for instance, states a feed conversion ratio of 12.1 compared to an FCR of 5.2 in Germany. However, as theoretical requirements per kilogram LWG are between 6.6 and 9.1 oat units for animals of 200 and 400 kg LW respectively, it would seem that the values of Germany refer to kilograms starch equivalents.

There is no evidence that there is a significant number of livestock farms without land; only few specialised landless beef fattening farms near big cities have been installed, but most feedlots are part of farms which also have land. With the exception of these few specialised beef farms, sufficient land for manure application is most probably available, but due to lack of economic incentives as well as large farm size, the distribution of the manure on the land is a problem. Probably the worst legacy of past collective policies is the massive environmental contamination caused by unsound farming practices.

The large livestock farms are sources of soil, water and air pollution. This pollution is exacerbated by imbalanced feeding resulting in a high nutrient excretion per kilogram production. Collectivisation and mechanisation has focused much attention on the use of grain and neglect of fodder crop and grassland production. However, ruminant livestock density in EE and the CIS hardly ever exceeds 10 LU per ha and a sizeable proportion of farm income comes from non-livestock activities.

In recent years, the livestock sector in EE and the CIS has contracted following removal of subsidies, increased prices of inputs and a reduced purchasing power of the population. Emphasis in livestock production is shifting from the large collective units to smaller units in the private sector. In the future more emphasis is necessary on fodder and pasture-based production and more balanced supplementary concentrate feeding. Grazing and a more even manure distribution over the available land will reduce environmental problems.

Urban Dairy Farming

Though (peri-) urban milk production is prevalent in nearly all developing countries, it is mainly important in South Asia particularly in India and Pakistan. An attractive market within the large urban population for such a perishable product like milk is one of the reasons that an urban production system has developed: it considerably shortens the distance between the milking animal and the consumer, thereby reducing the risk of spoilage of the milk and the marketing cost. Another major reason for the development of this system is the relatively low price of high quality by-products, compared to the price of roughage and milk.

Animals (mainly buffaloes) fresh in milk or just before calving are purchased from the rural areas and transported to the cities where they are kept under zero-grazing conditions. All feed is purchased, and at the end of lactation they are generally slaughtered. Unit size varies from around 10 to 100 lactating animals. Originally the units were spread all over the city, but governments have attempted to concentrate these units in so-called 'colonies' in the peri-urban areas, mainly for sanitary reasons.

Examples of this are the Landhi Colony in Karachi and the Aarey Milk Colony of Bombay. However, the ever expanding cities have again slowly absorbed such colonies into the urban area, and the inevitable growth of the cattle/buffalo population has resulted in overcrowding and is an environmental and sanitary hazard. Sometimes the peri-urban vegetable growers form a ready market for the solid manure, sometimes disposal of manure is a problem.

Urine and waste water is disposed of through the public sewage system or simply seeps into the ground water or the surface water. Van de Berg believes that the colonies rarely helped to solve the problem of the urban dairies. Many animals remained in the cities, legal measures were not fully implemented and the hygiene and manure disposal problems continued in the new colonies with high concentrations of livestock.

LLR system Production Trends

The statistical base for LLR systems is very weak. Total livestock population and production data are available per country, but separate data on the different production systems is not distinguishable. An additional problem is that the animals in LLR systems are bred and reared in other livestock systems so that only part of the total production of those animals can be attributed to LLR systems. The turnover in some subsystems within LLR systems is high and, thus, annual population data do not adequately reflect total production estimates.

Beef fattening in feedlots

Apart from the 'beef cycles', production and consumption has remained rather constant since the mid-seventies. This plateauing of beef consumption is owing to:

(1) income stagnation and population growth;

(2) reduction of the family income spent on meat;

(3) increased resistance to eating beef for health reasons; and

(4) competition from cheaper poultry and pork.

Poultry and pork production are expected to increase at sustained rate. These products will benefit most from the downward trend in cereal prices because they convert cereal into meat more efficiently than do ruminants.

In Canada and the USA the continuing shift towards specialised beef herds and an increased number of grain-fed animals, resulted in 1994 in a higher average carcass weight and higher total production.

Current estimates are that ca. 26 - 27 million head of cattle are fattened annually in the USA in approximately 10 million feedlot places. Fox assumes a production of 235 kg carcass per feedlot place. Another approach is ca. 1 Kg LWG per feedlot place per day or around 355 Kg per year equivalent to ca. 215 Kg carcass weight. Therefore, calculations are based on 10 million places with a production of 3.55 million tons LW. Around 84 % of the animals are fattened in large-scale landless feedlots. Around 75 % of the production is realised in the states Texas, Nebraska, Kansas, Colorado and California.

Veal production

Sere and Steinfeld estimated the veal production in OECD countries at 781,000 ton, representing less than 5.1 million heads slaughtered. This is lower than the number of calves slaughtered in the EU (Table 3), while also in the USA and Canada some 1.1 million veal calves are slaughtered. On the other hand, other types of calf production are included in the EU statistics. Calculations are, therefore, based on 6 million veal calves annually.

Table 3. Slaughter of calves in the EU 1992 - 19934. Slaughter of calves in the EU 1992 - 1993.

Country	*Calves*				*Carcass*	
	Slaughtered (000 heads)		*Weight (000 tons)*		*Carcass weight kg per animal*	
	'92	'93	'92	'93	'92	'93
Belgium	376	379	59	61	157	161
Germany	552	526	67	66	121	125
France	2376	2205	289	272	121	123
Italy	1514	1419	207	194	137	137
Netherlands	1197	1174	184	187	153	159
Other states	308	258	33	31	107	120
Total EU	6324	5962	839	811	133	136

The production of milk substitute fed veal calves is based on low prices for skimmed milk. With the present quota system to reduce milk production in the EU the production of veal calves has also reduced and a further decline is foreseen. Furthermore, public opposition to the production methods to produce white veal through iron deficient milk substitutes is growing. There is also particular antipathy towards the individual housing system for veal calves. Large-scale public protests in the UK have resulted in a halt to the export of bobby calves for veal production to mainland Europe.

Group housing is becoming increasingly common in the Netherlands and Germany and it is functioning well. In Italy and France scarcely any veal calves are housed in groups. In the UK group housing for veal calves is common, but the total veal production is limited. Research has also been done on individual housing, concerning the minimum box size from an ethological point of view and the limited supply of roughage.

Most of these changes are unattractive for veal producers, not because productivity will decrease but because the costs will substantially increase. These trends are more in line with the needs of the animals and the demands of society. Policies on animal welfare are largely based on the results of this research and are mainly focused on improving the health of calf group-housing systems. Prospects for the veal industry are determined by necessary investments in the environment and animal welfare. As this will reduce the profitability of the sector, a smaller scale production in the future is foresee.

Sheep Fattening

Sere and Steinfeld provide the only current estimates of the number of sheep fattened in feedlot situations. They estimate that on average 10 % of the sheep are fattened in feedlots, producing 15 % of the mutton in the area. Based on this estimate 10 million sheep are fattened annually in WANA, producing 100,000 tons of mutton, mainly concentrated in Iran (37%), Syria (21%) and Algeria (14%). These 10 million sheep are only fattened for ca. 70-120 days. There are indications that in some countries a higher percentage of the sheep are fattened. However, it is not clear if the data refer to peak periods or to annual averages.

Little is known about intensive sheep fattening trends in WANA. Sere and Steinfeld estimate annual growth rates for the period 1982 - 1992 at almost 10%. Increasing incomes and population growth has led to an increase in the demand for mutton. The demand for mutton has been met in the past by animals from the pastoral areas. Fear of overgrazing and degradation as a result of increased stocking rates has led to promotion of intensification through supplement.

Barley is the main feed supplement. Feed subsidies are common even though nearly all WANA countries are net importers of barley. In Jordan, for example, the government feed subsidy is 35%. The main effects of these subsidies are a growing independence of sheep farming from rangelands, a higher degree of market orientation and a further sedentarisation of sheep farmers. In most WANA countries these barley subsidies are or have been reduced, but the price ratio feed/mutton remains attractive for intensive fattening.

Large-scale Beef Production

Accurate estimates of livestock numbers in EE and the CIS are scarce, and the proportion of ruminants in landless livestock system is completely unknown. Sere and Steinfeld estimated that 40 % of the livestock population would be part of LLR systems, producing also 40% of the beef. This estimate is rather weak as no dairy farming is included while growth rates are assumed to be similar to land based livestock production. This estimate assumes LLR systems to be far more important than data from other sources suggest. According to Mudakar et al. 1992, the cattle population in CIS, by far the most important region, consists for 3 % of beef breeds and 37 % of dual purpose cattle. It is likely that not all animals are kept in LLR systems. Also, solely in the CIS, it has been indicated that specialised beef fattening farms were producing less than 4% of all beef, while part of these farms were not landless or did not have sufficient land in the near surroundings.

It has been argued before that, strictly speaking, the LLR system hardly exists in EE and the CIS, but that their large-scale livestock units could have comparable problems. Data on the proportion of livestock in the private, the state and collective sector could give an indication of the extent of the problems. In the CIS in 1990, for example, 79.6 % of beef cattle were kept at state and collective farms producing 74.5% of the total beef production. However, the data is incomplete and varies from country to country, partly due to recent political developments. Moreover, these developments also make the assumption on the less concentrated nature of private farms untenable, as in some EE and CIS countries privatised state farms have not been abolished but continue to produce in a similar way as the former state farms.

Urban Dairy Farming

No statistical information is available on urban dairy farming. Nestel estimates around 1 dairy cow for every 10 citizens in urban areas in India, with city milkers commonly owning 10 to 20 cows. State Governments are attempting to shift these urban dairy units to rural areas, the Aarey Milk colony of Bombay is an example of this. Maki-Hokkonen estimates 260,000 dairy animals in Karachi in Pakistan producing daily 1.7 million litres of milk. The average herd size is around 55 animals; within the Landhi colony the herd size is around 100 head. The human population of Karachi is around 6 million.

Sere and Steinfeld did not include this production system in their estimates, because they assumed that hardly any manure problems exist. However, as concluded earlier, manure problems do exist partly because the lower DM content of the manure as a result of high concentrate rations makes this manure unsuitable for use as fuel. Moreover, little land is available for manure application in the surroundings of these farms due to their location in peri-urban areas.

Motives

Though LLR systems are very heterogeneous, most subsystems seem to have one feature in common: a high livestock product/concentrate price ratio. In the USA, for example, live weight/maize price ratio was ca. 13-14 in 1992. In the EU beef fattening on high concentrate rations is less common probably due to a lower price ratio of 9-11. In most developing countries this ratio is much lower (2-3) making feedlots generally uneconomic, though they exist in some countries, mainly in (small) areas with a good supply of cheap concentrates but sometimes also due to political incentives.

The major exception are the LLR systems in the CIS and EE where price ratios were less favourable, though this picture is very confusing as manufactured feed was subsidized. These less favourable price ratios did not prohibit high concentrate feed rations, partly because political incentives were overruling economic logic: state and collective farms were not driven by the need for profitable livestock production as farm gate price of cattle in Russia in 1992 covered only 60-73% of the cost of production. Hence these farms incurred major debts: highly problematic in the economic adjustment period. It is envisaged that the practice of high(er) grain utilisation for feed will gradually disappear in the near future mainly due to privatisation of large collective farms and economic adjustments.

In the case of veal production and sheep fattening, subsidies on milk and barley play a crucial role. An extreme example seems to be Algeria where the mutton/barley price ratio increased from 30 in 1970 to 66 in 1987. Also in other WANA countries barley subsidies in combination with a free market for mutton has increased mutton/barley price ratio considerably. Favourable price ratios are as a result of low concentrate prices and in many cases also the result of the high premiums paid for the specific quality meat produced by the LLR subsystems. In developing countries usually no premiums are paid, so LLR systems in most of them are not economically viable.

4

Livestock Production in Mixed Farming Systems

As rural population pressure increases and less land becomes available, both crop and livestock farmers need to intensify. McIntire *et al* showed that as population pressure increases, the two activities often integrate. The soil protection techniques, and in particular the long fallow (rest) periods under traditional cropping systems, which allow soil nutrients to regenerate and soils to be protected from erosion are not sufficient anymore.If farmers can not resort to external inputs, the integration of livestock and crop activities represent their main avenue for intensification. Mixed farming has thus constituted the basis for modern agriculture.

Mixed farming systems provide farmers with an opportunity to diversify risk from single crop production, to use labor more efficiently, to have a source of cash for purchasing farm inputs and to add value to crop or crop by-products. Blending crops and livestock has the potential to maintain ecosystem function and health and help prevent agricultural systems from becoming too brittle, or over connected, by promoting greater bio-diversity, and therefore increased capability to absorb shocks to the natural resource base. Mixed farming systems:

1) maintain soil fertility by recycling soil nutrients and by allowing the introduction and use of rotations between various crops, forage legumes and trees or land to remain fallow whereby grasses and shrubs become re-established;
2) maintain soil bio-diversity and minimise soil erosion, water conservation and provide suitable habitats for birds;
3) provide the best of using crop residues. If the stalks are incorporated directly into the soil, they act as a nitrogen trap, exacerbating deficiencies. Burning, the other alternative, increases carbon dioxide emissions.

If pressure increases further, the crop-livestock systems can decompose into specialised crop or livestock activities. If there are no improved market opportunities, such as the case in many developing countries, and as human population pressures builds further, increased rates of nutrient depletion and erosion easily follow in the arable land part of the system. This can in turn lead to an involution or downward spiral of mono-culture with lower quality food crops, increased under-nutrition and famine. However, if urban incomes rise, more market opportunities open up and farmers become more integrated into the market economy, specialisation, allowing farmers to take advantage of economies of scale and develop greater levels of expertise, normally occurs. Finally, under very strong demand, and with high input or price subsidies, as such exist in many developed countries and in the fast growing economies of East Asia, excessive importation of nutrients can lead to soil and water tables being overloaded with nitrogen or phosphorus.

The challenge for the mixed farm sector will therefore be to maintain a nutrient and energy equilibrium through crop-livestock integration and at the same time allow sustainable productivity growth. While the mixed farming system, more than in any other production system, operates under a wide range of environmental and economic conditions, requiring regional solutions and practices, there is one overriding criteria in determining the size and nature of the system's impact on the environment. This is the nutrient balance. This balance is determined by the nutrients (N, P and K) brought into the farming systems by in-organic fertiliser, feed, Nitrogen fixed by leguminous plants, transfer from - mostly common - grazing areas outside the farm, and the amounts exported in animal products or lost from the land in the air or ground water.

A positive balance will have completely different effect than a nutrient deficient system. In this analysis, the mixed farming systems will therefore be classified in nutrient deficient systems, mostly occurring in the developing world, and of nutrient surplus systems, mostly found in the industrialised world and increasingly in the fast growing economies of East Asia.

Nutrient Deficit Systems

Like in the grazing systems, human population pressure, poverty and poor infrastructure are the most important fundamental driving forces affecting the environmental impact of mixed farming systems in the developing world. And in that respect, the mixed farming systems are entering a dynamic period of growth and change, as they cover some of the countries with the highest birth rate in the world. Human population pressure affects the critical nutrient balance. In many systems in Africa and the mountainous areas of Asia and Latin America, this balance is being maintained with - at best - small inputs from outside, as poverty and poor infrastructure often preclude this balance to be maintained with outside feeds and fertiliser.

The balance then depends mostly on the ratio between crop and non-arable land. In the case of Nepal, it was determined that 3 ha of forest land with livestock are needed to maintain 1 ha of arable land, without causing deterioration of the forest. In West Africa, the ratio was about 10 – 40 ha of dry season grazing and 3 – 10 ha of wet season grazing to provide adequate nutrients for one ha of millet crop land.

However, with increasing population pressure more and more grazing land is converted into crop land, the crop/grazing land ratio narrows, and the nutrient flow into cropland decreases. For example, cropland increased in the Machacos area of Kenya from 35 percent of the higher potential area in 1948 to 81 percent now. In Northern Mali, the rate of increase of the arable cropping area was the same as the 2.6 percent rate of annual population growth over the last twenty years. This can lead to increased competition for land and grazing on the open rangelands, which in turn, can lead to privatisation of crop residues and rangelands.

For example, the common property lands in India, decreased since 1960 by 50 percent, so that in 1992 they constituted only 15 percent of the total area. Population pressure causes farmers to push cultivation out into more resource limited areas, causing soil erosion.

Mabbutt reports, for example, in Niger millet fields are appearing 100 km north of the recommended limit to cultivation, increasing the risk of crop failure and leaving soil exposed and erodible, and as a result livestock have been moved to the most marginal areas. With even further deteriorating crop/grazing land ratios and without adequate substitution of soil fertility by other sources, fertility gaps arise. This is typically the case of many mixed farming systems in the tropics.

Reported losses range from about 15 kg N/ha/year in Mali, to more than 100 kg N/ha/year in the highlands of Ethiopia. Downward spirals of declining soil fertility, overgrazing, increased erosion and losses in soil micro flora and fauna are the result and will ultimately cause significant decreases in agricultural productivity and political stability. Events in Rwanda underscores the need for a balance between crop activities, animal inputs and human population pressure, to avoid the involution of farming systems, impoverishment and even civil war.

In irrigated mixed farming system, grazing is more contained on the farm. In these areas, the type of feed supply will shift increasingly from non-arable grazing to grazing crop-aftermath and utilisation of fallow land. Soil nutrient deficits are covered in these areas by inorganic fertiliser. The main feature of the irrigated areas has been the dramatic intensification of crop production as a result of the Green Revolution.

With the introduction of improved "dwarf" varieties, the quality and quantity of crop residues decreased, but cereal production and total farm income increased. This has decreased the amount of feed available for ruminants, but increased the supply of cereals, which could be used for intensive production. Moreover, the rising incomes from the Green Revolution increased demand for livestock products. The two trends together promoted the development of intensive industrial units in these areas. This, combined with the high fertiliser use, leads to increasing concerns for nutrient loading.

Pressure is also on the use of draught power, the other link between crops and livestock production. Off-farm employment increases labor cost and if the time available to prepare land becomes shorter, tractorisation becomes more important, decreasing buffalo and cattle populations for traction. For example, in India the number of bullocks used for animal power decreased from 85 million in 1960 to 60 million now, at the same time the number of tractors increased from 30,000 to 1.4 million. Only about 30 percent of the total arable area is cultivated by animal power.

Nutrient Surplus Systems

Soil erosion increases with increasing levels of grazing pressure and cropping intensity. Reynolds et al. reported how variable soil erosion can be observed under different grazing pressures. Soil loss on lands with good ground grass cover was estimated at 1 ton/ha; this increased dramatically on overgrazed pastures to a level of 53 tons/ha. If grazing pressure is in balance with the forage resource being produced, livestock's interaction can encourage more stable land-use practices.

Pimental et al. discussed how in the United States in order to increase farm size, grass strips and shelter belts were removed, thus increasing the erosion rate. In the past, livestock provided a rationale for grass strips and shelter belts, but with the disaggregation of crop and livestock systems in OECD countries the rationale for utilising land resources in such a way has decreased.

Nutrient surpluses cause the main pressure on the quality of land and water. These surpluses come from a combination of inorganic and organic fertilisers. Inorganic fertiliser is used, in spite of the fact that the system could be balanced based on the nutrients supplied by the manure produced in the systems. For example, in the case of Brittany, the organic manure produced by the intensive industrial and mixed farms of the region would adequately balance the regional nutrient balance, but large surpluses emerge because of the additional importation of large quantities of inorganic fertilisers.

The inorganic fertiliser is said to be necessary because of the readily available nutrients and its easy transport, as in the case of pure intensive grazing systems. The preparation of nutrient balances (inorganic and organic) on a regional basis is therefore a critical element in identifying land conservation and fertiliser policies.

For a large part, the heavy import protection of most animal products, especially beef and milk, and the wide array of producer subsidies, especially on feed imports, have led to the wide use of feed concentrates and the surplus nutrient situations.

Technology and Policy Initiatives

A fundamental requirement to achieve environmental sustainability of the natural resource base is to incorporate change and flexibility into the respective production systems and overall landscape. Agriculture and particular crop agriculture seeks to stabilise production and environmental change, causing a loss in ecosystem flexibility. But the vibrancy of any ecosystem which supports the production of food, can be enhanced through the integration of livestock as a mechanism to promote system flexibility.

Mixed farming, with careful attention to the nutrient balance to prevent system involution, is probably therefore the environmentally most desirable system, and should be the prime focus of agricultural planners and decision makers. The challenge will be to identify the technologies and policies, which generate the sustained and accelerated growth needed to satisfy the world's sky-rocketing demand for meat and milk. This challenge will be enormous. Past global growth of the meat and milk output of the mixed system has remained behind the increase in global demand.

Will the growth in demand levels continue to rise beyond the capacity of mixed farm systems to meet national demands and therefore result in an immediate transition to highly intensive industrial systems? This study argues that, while for environmental reasons establishing the appropriate enabling environment for mixed farming is critical, mixed farming should make an important contribution to meeting global demand, it might not be enough, and industrial production growth will be essential.

The key enabling factor seems to be access to inputs and attractive markets, as illustrated by the Machacos case study in Kenya. Here, a dynamic market and access to inputs led to sustainable development of a resource poor area, of which the population had quintupled over the last sixty years. Use of livestock in such intensive situations includes a shift in species from cattle to sheep and/or goats.

Driving this option would be the access to markets and the infrastructure necessary to process milk. Farmers with more difficult access to markets would be more inclined to raise small ruminants. This path has the biggest potential in the highlands areas of Latin America and the semi-arid and highlands of sub-Saharan Africa, where the land to human ratio is already rather high and integrated crop-livestock systems have an immense potential to contribute to the required growth in productivity.

The second development path could lead to specialisation in either crop or livestock production. Controlling this path will be existing infrastructure, and relative price ratios between the different inputs and outputs and economies of scale, which will determine the comparative advantage in one enterprise versus another. This is to a large extent the development path in the developed world and East Asia and will be discussed separately below.

In mixed farming systems, there are exciting opportunities for introducing technologies. Applied research and extension are of critical importance to maintain the environmentally friendly factors of the system, although McIntire *et al.*, strongly make the point that African farmers are well aware of the technologies involved. The challenge will be to convince farmers about technology's utility. In the nutrient deficient systems of the developing world, the emphasis would be on control of soil erosion and improving nutrient recycling. Some examples are:

i) Improvement of soil cover through the use of alternative crops for mulching, and introduce soil management techniques such as conservation tillage, bench terracing, strip cropping, contour farming, etc.;

ii) Improvement of feed production and quality to reduce the pressure on the grazing areas, and improve internal nutrient transfers. Technologies to do so include:

— introduction of fodder shrubs and trees planted to reduce soil erosion and improve soil fertility. Several mixed farming systems using fodder shrubs and trees have been developed. An example is the agro-forestry systems with three strata as successfully introduced in Indonesia;

- — improvement in feed quality, for example through urea treatment of feeds, as reportedly successfully used now by more than 2 million farmers in China and through increased efforts in plant breeding to correct the grain bias of the green revolution "dwarf" varieties and raise nutrient quality of the crop residues;
- — use of non-conventional feeds;

iii) Reduce nutrient losses from manure and improve the efficiency of their application by:

- — promotion of stall feeding which doubles the effective availability of Nitrogen and Phosphorus;
- — strategic supplementation for specific classes of animals to improve the efficiency of limited amounts of available feed.

iv) Increase production efficiency, and thereby farm income, resulting in improved purchasing power for soil improvement and conservation methods. They include:

- — within breed improvement;
- — where appropriate changing to small ruminant production or to optimal combination of ruminant and non-ruminant species;
- — strategic supplementation of lactating and growing animals.

In the developed world, strong regulatory frameworks are being introduced, restricting the emission of nutrients in the case of point-source pollution, and stocking rate restrictions in the case of non-point source polluters. They include a variety of restrictions on stocking rate, level of fertiliser, manure storage, application techniques and times, along with government subsidies on manure processing and management.There are a variety of incentive instruments to reduce nutrient surpluses in nutrient surplus areas. These include:

— the taxation of inorganic fertiliser, to reduce the incentives for adding inorganic nutrients to an already surplus system;

— maximum application limits, and regulation of the time of application, to reduce leaching and volatilisation;

— taxation on feed imports to reduce the significant nutrient transfer from the already nutrient deficient developing world, to the nutrient surplus of the industrialised world. Such taxation might be supplemented with or used for subsidies for a transfer of nutrients back to the developing world;

— incentives to achieve a more balanced distribution of crop and livestock activities.

Technologies to reduce nutrient surpluses are also described under the industrial systems. But of particular importance to mixed systems are the following:

— Improved nutrient formulation in terms of incorporating optimum levels of Nitrogen into feed and adding enzymes to improve the utilisation rate of plant Phosphorus. This optimum nutrient ratio and composition management reduces the loading of the environment with these elements;

— Better manure storage and application of manure. Improved manure handling and storage are a key requirement, for a more efficient use in the nutrient deficient areas, and to prevent leaching and pollution of surface water in the nutrient surplus systems.

In addition, over the last decade, there has been a considerable interest in promoting low-input mixed farming systems, as sustainable, and environmentally friendly systems. In the USA, the Rodale Institute in Pennsylvania, and the LISA (Low Input Sustainable Agriculture) movement, strongly promoted by USDA has been in the forefront. In Europe, the ILEA (Institute for Low External Input Agriculture) has been one of the prime movers. The mixed farming system, with its rather closed character, is especially suitable for this type of production. Some of the main technologies, which can be used on the livestock side are:

- — Use of grass-legume mixtures as the basic feeding system for ruminants. The Nitrogen fixation of the legumes, can replace the Nitrogen lost through milk and meat as well as serving as a sink for excess Nitrogen;
- — Mixed species grazing which increases grass yield, botanical composition and reduces parasite load;
- — Genetic improvement focusing on maximum capacity to digest roughage, and based on life-time production;
- — Housing, which allows adequate movement and is based on the animals natural social behavior, produces high quality manure using large quantities of straw;
- — Disease control based on good feeding and breeding practices, and preferably using herbal and organic medicines to minimise environmental contamination.

This has to be accompanied by environmentally friendly low-input techniques, such as no-till, manuring and composting, integrated pest management etc. The opportunity to employ precession agriculture concepts, due to the information available concerning the resources base and the economic incentive, to minimise production costs provides a mechanism to keep soil nutrients balanced as well as mosaics across a landscape.

5

Urban Livestock Production

An urban livestock system is characterised by a large variation of livestock systems that occur in and around densely populated areas and that strongly interact with the surrounding wealthy as well as poor human communities in different ways, at several levels of system-hierarchy and with nearby and distant rural areas.

Livestock raising has often been part and parcel of urban agriculture, presenting its own specific problems and opportunities. Until recently, it was often regarded as problematic, backward and a sign of poverty. As with all branches of urban agriculture, however, livestock keeping now seems to be recognised for the positive role that it can play in urban living conditions across the world. Indeed, livestock production has a variable and controversial, but often essential, role to play in and for cities. It occurs on a small scale, with both small and large animals. It is subject to change, e.g. local breeds are replaced by foreign breeds .and, while it is increasing in some cities, it is decreasing elsewhere.

Urban consumers use more animal products than rural consumers, and consumption as a whole seems to be increasing. Neglect of the livestock sector can only lead to negative developments, while positive attention can help to uncover the opportunities inherent to this form of livestock production.

History and Development of Urban Livestock Systems

All cities differ from one another in size, development history, sociology and structure. Mexico City has a population of around 22 million people, while Abidjan (Côte d'Ivoire) and Dar-es-Salaam (Tanzania) count about 2 and 3 million people, respect ively. When thinking in terms of future developments in opportunities for urban livestock, distinctions can also be made between cities that have grown in blanket form and cities that were planned according to the concept of satellite towns.

Obviously, a range of intermediate forms exists between blanket and satellite patterns, and practice does not always follow planning. Some cities opt for decentralised planning or *laissez faire*, others for centralised planning, and still others seem to have developed without any clear plans at all. Singapore is a typical case of a city with strong legislative measures, but in other cities hardly any effect is expected from legislation other than an increase in bribery.

Keeping dairy cows may be impossible in the heart of modern Mumbai (India), Tokyo (Japan) or Amsterdam (the Netherlands); but rabbits, chickens and other small (pet) animals are commonly kept on balconies and in backyards everywhere. Even the keeping of dairy cows and/or goats can be quite acceptable in urban areas with much green, or where there are a lot of by-products from agro-industries. In the nineteenth century, urban dairies were quite important in areas around beer breweries in Copenhagen (Denmark) and around distilleries in major United States cities, while today they still supply a good income in Dar-es-Salaam, particularly to civil servants.

Another potential criterion for distinguishing among urban livestock systems is in their origins. A system may emerge through:

— "organic" emergence, i.e. the system evolved by itself;
— induced development, often with public money;
— city expansion, e.g. when previously existing village farms are swallowed up by urban development to create overgrown farming;
— importation of livestock by rural people immigrating into the city;
— traditional tribal groupings that settle in towns while maintaining their liking for animals;
— situations in which livestock is seen as an emerging opportunity for cleaning waste
— cities with wealthy people where animals can provide emotional attachment or a learning opportunity.

Husbandary Techniques

A traditional approach to animal production focuses on what is called "disciplinary" aspects, such as animal feeding, breeding, health, housing, animal behaviour, farm economics and product processing.

Feeding Strategies

Feeding strategies in urban areas vary according to such factors as social categories, animal species, household income and distance to the city centre. An important difference between animals is the distinction between ruminants such as cows, buffaloes, goats and sheep, which eat feeds rich in fibre (grass, hay, straw), and monogastrics such as chickens, ducks and pigs. Resource-rich livestock-keeping families feed their animals to reach certain production levels and can buy as much feed as they need. Resource-poor farmers make the best of what they have. They lack the financial means to buy commercial feed and the labour availability to collect or cut feed while producing fewer household residues.

In most urban areas, roadsides and empty plots on the outskirts are a source of cut grass, and are also used to produce green fodder (pigeon pea, alfalfa); in addition, cereal or vegetable residues are sometimes used. Fodder markets import feed from outside, especially in more centrally situated areas that have limited space for fodder production.

Although even in the most built-up areas it is possible to establish small plots (gardens), and even roofs can be used, in urban conditions green feeds and fibrous crop residues tend to be relatively scarce. When (green) fodder is scarce, farmers can either turn to raising smaller animals (goats or sheep instead of cows or buffaloes) or buy concentrates (rice bran, oilseed cakes, commercial meals).

By-products such as kitchen wastes, hotel leftovers or waste from the canning industry and beer brewers can also be useful, even in the keeping of such large ruminants as cows and buffaloes. All of this puts the animals into a useful ecological niche: that of cleaners of waste products.

Animals help to reduce the organic matter of the waste and they do not add a particular nutrient load unless large amounts of feed are imported. The scale of animal keeping is, of course, related to the amounts and kinds of feed available; when feeds for animal production are imported, nutrient surpluses can occur, leading to environmental problems.

Depending on the animal species and type of production involved, three major feeding systems can be distinguished. These are based on the use of household wastes and agro-industrial by-products, roughage and concentrates, respectively.

Systems based on household waste and agro-industrial by-products

Poultry, pigs, rabbits, guinea pigs, pigeons and even fish in small units are mostly fed household wastes. These units can be found everywhere: in city centres, in backyards, on balconies and on rooftops. Larger units use industrial by-products and commercial feeds. UAM estimates that 4 000 tonnes of food wastes are used every day in Mexico City alone. These farmers collect wastes from bakeries, restaurants and the kitchens of communal institutions, just as Western European farmers did in recent times. The wastes used include kitchen wastes, stale bread and tortilla, leftover tortilla dough, chicken guts, and fruit and vegetable wastes from the markets.

It is estimated that 100 tonnes of fruit and vegetable wastes are used by approximately 2 500 cows, producing 37 500 litres of milk a day. These wastes include maize husks, lettuce, cabbage, cauliflower and radish and broccoli foliage; carrots, beetroots, maize and Castillean squash that are unsuitable for human consumption are also available.

Important waste products for pigs and dairy animals are brewers' grain and rejected bananas or pineapples from canning factories. Small units of such species as rabbits can be fed diets that are based entirely on wastes and some roughage. Units in Sokoto have between 3 and 19 animals, 60 percent of them having fewer than ten. These rabbits are fed on garden trimmings and leftovers, kitchen wastes, grains, lettuce, carrot leaves, other leafy vegetables and Amaranthus cultivated on small plots.

Roughage-based systems

Feeding of large herbivorous species (sheep, goats, cattle, buffaloes) in urban areas is complex, because a large proportion of their ration has to be fibre in order to ensure a good functioning of the digestive system. Roughage such as straws and grasses contains a lot of fibre, but in urban conditions these feeds are generally expensive. They are rarely produced inside cities, and the costs of transport and storage of bulky fodders are high.However, urban production systems using expensive roughage can be profitable in particular circumstances, including:

— In cities, goat, cattle or buffalo husbandry is almost always related to dairy production. If there is a demand for fresh milk, feeding milk-producing cows and their calves with industrial by-products and expensive roughage is economically feasible. Even large commercial production units are remunerative. In India, the high demand for fresh buffalo milk leads to large commercial units of up to 500 lactating buffaloes inside cities. Fodder is bought from specialised grass producers and cutters living on the city outskirts, who often irrigate (treat) the grass they produce with sewage water. Dairy farmers are prepared to pay high prices for these feeds and grass is becoming a cash crop for local smallholders. Semi-zero grazing systems develop when grazing opportunities increase, e.g. on roadsides. Zero-grazing systems seem to be labour-intensive but, as units are small, the cutting and carrying of feed in urban conditions is less time-consuming than herding.

— In countries with a high proportion of Muslims, uncastrated male sheep are slaughtered at such religious festivals as Ramadan and Id. Animals that are raised elsewhere are brought to the city to gain a lot of weight in a short time. In Mali, the main feed ingredient for these animals is cereal straw complemented with industrial by-products such as oilseed, groundnut and cotton cake and cereal bran. Feed may be expensive, because the market prices of these animals are high.

The roughage needs of small species such as rabbits can easily be satisfied, particularly if they are kept in small numbers. Moreover, rabbits, fish, birds, pigs and other omnivorous species need very little roughage, although pigs can digest diets containing a considerable proportion of fibre. Vegetable waste is very useful for pigs, provided that the growth rates and meat gradings aimed at are not based on other production circumstances.

Concentrate-based systems

These systems concern mainly intensive and specialised poultry, pig and dairy units in peri-urban areas. The concentrates are often imported or made from local grain and oilseed milling by-products. Locally produced commercial feeds are often too expensive to serve as full ration feeds, and adulteration is common.

As a result poultry, pig and dairy farmers in some regions tend to use concentrates as supplements to rations based on waste products, thereby focusing on economic issues rather than nutrient-use efficiency. Research and extension could be made more relevant to this by studying the economic effect of different waste-product mixtures rather than concentrating on ways to achieve biological maximum levels.

Whereas in Western countries much attention is paid to increased nitrogen balances and standard economic efficiencies, in tropical conditions it is important to understand the economics of small versus large farmers. Moreover, the problem of nutrient disposal in urban conditions probably has little to do with nitrogen or phosphate concentrations, and more with the periods when and the form in which dung becomes available: chicken dung can be fairly dry and easy to manage; pig dung tends to be thin; and cow dung can sometimes be separated into solid and liquid fractions. Other issues of concentrate-based rations to be addressed are the occurrence of micotoxins and chemical residues in the feed. Where legislation is ineffective there seems to be little point in establishing laboratories; collective action and education will have to be part and parcel of legislative measures.

Feed efficiency

There is a tendency to think that high production levels go together with a high feed efficiency. However, feed efficiency can be defined in different ways. If all inputs (breed, feed, housing, veterinary care) are adapted to a high production level (for instance fast growth), even the use of such concentrates as grains that are also valuable for human consumption may be economically efficient if the farmer is rich and if there is a cereal surplus.

In other circumstances, for instance if cereals are imported and many farmers are poor, the use of expensive cereals as concentrate, expensive housing, veterinary care, etc. might soon cause greater farmer poverty and accelerated cereal price increases, leading to a declining demand for meat and decreasing revenues. Maximum efficiency will only be attained when locally available resources are used as much as possible by as many as possible resource-poor local producers, even if production levels are low. When even relatively low gains on otherwise useless "waste" are attained, high feed efficiencies can be achieved and live weight gain is obtained without, or with very low, concentrate costs.

Live weight gain increases with supplementation, but feed becomes more expensive. If no concentrates are fed, the margin (benefits minus feed costs) is positive when the live weight value exceeds SL R 7.3/kg; if concentrates are fed, the margin is positive when the live weight value is more than 12.6 SL R/kg. If, instead of roughage, only waste is fed, the margin is always positive as long as the animals gain weight without supplementation; with supplementation the margin becomes positive when the value of live weight exceeds 10 SL R/kg. In this respect, the reorientation of planners and technical staff can bring about great changes in government policies, resource use efficiency and the redirection of research and development priorities so that they allocate all funds and attention to rapid high gains rather than to efficient production.

Reproduction

An efficient reproductive performance is important for the sustainability of animal production systems. It is essential for the provision of replacement animals and, in the case of dairy animals, the regular production of calves is also necessary for milk production. In cities, however, feed is expensive, space is limited and, as a result, any reproduction that does not generate immediate revenues represents an additional management burden. Simple solutions include the "externalisation" of reproduction, under which dry cows are taken to the countryside to conceive and calve again and poultry breeding takes place even further away.

Efforts to make stressed city animals conceive through the use of hormones are of limited benefit: they may be successful but tend to exhaust cows more quickly, making them fit only for early slaughter. More efficient production can be achieved when breeding takes place in better climates or in places where land and labour are cheaper. One way of saving resources and securing foreign genes is artificial insemination. In urban conditions, the importance of reproductive management varies depending largely on the types of animals involved and the scale of inputs and outputs.

Rabbit reproduction is essentially an on-farm activity. Local breeds can be improved through selection of the best local animals or through selective distribution of breeding animals from elsewhere. Commercial poultry gets its breeding stock from specialised and highly commercial units, and has little scope for local breeding programmes.

Backyard chickens, as with rabbits, can take care of their own reproduction, and are a good point of entry for the establishment of local breeders' groups; an approach that can also be taken to the breeding of goats and sheep. Cow or buffalo operations tend to obtain young stock from rather informal markets in the countryside, which are often hundreds or even thousands of kilometres away. Breeding policies for such animals should not be directed at urban farms but at the source of cattle supply.

Whether in cities or villages, poor or rich families, the choice of breeds should be adapted to local conditions in order to obtain optimal results. Imported European high-performance breeds (which require expensive concentrate feeds and special housing), when introduced to local conditions, will often produce even less than local animals. Part of the reproduction process is the rearing of young animals. Again, this is often done in the countryside or even in other countries. Typically, good results can be achieved by paying attention to hygiene, particularly when small animals such as rabbits are concerned.

Breeding

When local breeding is concerned, it is important to distinguish between natural and artificial breeding. There are two types of natural breeding: free mating in the range or controlled mating. In the former system, male rabbits detect when a female is on heat, and females are usually mated several times during each heat period. One male can cover 40 to 50 cows per year, provided there is no marked seasonality in the occurrence of heat. In controlled mating, it is the farmer who carries out heat detection and decides when matings should occur.In these cases, a doe is mated once or twice during each heat period, and a buck can be used on three to four does per week or 150 to 200 does per year.

Free mating in rabbits can be disastrous because the females bear too many litters, become exhausted very quickly and produce weak kittens. Mating at too young an age, for example in goats and sheep, is also detrimental to the future performance of the animal. Some form of mating control is therefore essential.

Artificial insemination (AI) is the most important reproductive technology in many livestock production systems in temperate and tropical regions. With AI, one ejaculate from a bull can be used to serve 400 to 500 cows and, under optimal circumstances, one bull can produce sufficient semen for more than 50 000 cows a year. Advantages of AI are that farmers do not have to undergo the costs or hazards of rearing breeding bulls, and have access to a very wide range of bulls. Furthermore, many infectious reproductive diseases can be controlled by the use of AI. On the other hand, the disadvantages of AI include high overhead costs for centralised breeding centres, equipment, personnel and training. AI also requires good infrastructure, semen distribution, field inseminators and, if frozen semen is used, a regular supply of liquid nitrogen. Farmers must also be educated in heat detection and the timing of service, and a reliable system of communication with the AI service has to be in place. AI is especially suitable for livestock in peri-urban areas, if at all, in developing countries.

Reproductive problems in urban livestock are quite common as a result of production stress, a shortage of male animals, poor management on the part of farmers, etc. Many methods are available to detect reproductive disorders, including improved farmer skills and hormone injection. More modern reproductive technologies are also available to livestock breeders. However, only a few of these technologies can be applied to tropical (smallholder) dairy systems, AI being an example that is used in certain areas. Embryo transfer (ET) is a still more exotic technique that involves the use of certain females as donors of embryos that are then implanted in recipients (surrogate mothers). This allows the preferred animals to produce many more calves during their lifetimes than would be possible by natural means.

Heat detection itself is difficult in tropical stress situations and it requires close observation. More recently, heat can be detected by measuring the concentration of a hormone such as progesterone. Small amounts of the hormone are present in blood and milk, and techniques such as radio-immunoassay and enzyme-immunoassay enable these to be observed closely for heat. Animals can then be mated at the correct time. This process is also used in pregnancy diagnosis, a method that requires skilled technicians.

Management of reproduction

Records are an indispensable component of modern farming management. Carefully recorded observations avoid the likelihood of supporting unproductive animals with such inputs as feed and housing. Practical farmers often remember important management information, even if they do not write it down. However, failure to keep records can lead to the loss of important information.

In addition, farmers may not be able to identify heat signs, recognise disease sufficiently well, etc., so it is useful to encourage that at least a simple system of identification and recording be followed. Women play an important role in this activity, and the least a farmer can do is to maintain a record of calvings and dates of littering/farrowing, vaccinations, etc.

Other important aspects of management are:

— hygiene in the rearing of young animals;

— the timely treatment of wounds;

— provision of adequate drinking-water;

— alleviation of heat stress through grazing management and design of housing;

— control of parasites and diseases;

— keen observation of (re)productive status.

Chickens can easily be selected by checking whether they are actually laying or not. Early detection of disease and disorder is essential. Most important may be the choice of animals; genetic characteristics of the animals must be compatible with the environment and the available resources, including the knowledge and skills of the farmers.

Animal Health and Food Safety

In urban livestock systems, the close association of people and animals causes special risks in terms of zoonoses. These are diseases that affect both animals and humans, for example, parasites, tuberculosis, brucellosis and anthrax. Furthermore, special feeding conditions and high animal densities may cause special metabolic disorders (acid rumens) and animal welfare problems.

African Swine Fever

This is a virulent tick-borne disease in pig breeds which was introduced to Africa hundreds of years ago. The virus naturally exists in warthogs, which are little affected, and in a tick that lives in the burrows of these wild pigs. The virus is spread to other regions by living animals, meat, meat products, swill and dung. In recent years the virus was introduced into West Africa where it caused catastrophic losses among domestic pigs in countries such as Benin, Côte d'Ivoire, Ghana, Nigeria and Togo. The disease has become endemic in parts of West Africa, Latin America and Europe. In new unexposed pig populations the infection is fierce, killing the majority of the infected pigs.

African swine fever is a "List A disease" of the International Office of Epizootics. Infected countries are expected to take measures to control the disease. However, no vaccine exists and the only way of combating the disease is to kill all the animals in the affected region. This approach is usually bound to fail, as it is not in the interest of (poor) pig keepers who hide their animals or otherwise fail to cooperate with this kind of measure.

Indeed, many backyard pig breeders live in shanty towns, where veterinary control is often repressive rather than being a service. In addition, undertakings to compensate for destroyed pigs are often not kept. Ideally the decision to stamp out the disease through wide-scale slaughter should be taken democratically and the population should be compensated for losses.

Cisticercosis

Cisticercosis is a tapeworm infection (Taenia solium) that passes from animals to human beings. It is prevalent in both rural and urban areas and is closely related to the economic standard, aspects of hygiene and the way animals and people share the same living space. Major problems with this disease occur in Latin America and non-Islamic parts of Africa and Asia, especially India. In heavily infected animals, small cisticercus cysts, which are the larvae of the tapeworm, can be felt in the pig's tongue. The population has no general understanding of how pigs get the cysts, but in places where people know about the disease, the price that can be fetched by such an affected pig is considerably lower than that for a pig without cysts.

In some regions, between 15 and 60 percent of the pigs kept in traditional systems have cisticercosis. Pigs become infected when they eat human excrement. Free-roaming pigs, particularly when there is a lack of latrines, are important in the reproductive cycle of the parasite. Infections do not occur in intensive pig keeping, where the animals are raised in an enclosed space and cannot eat human excrement. People get tapeworms when they eat uncooked meat that contains cysts. In a population, the percentage of people harbouring the Taenia is an indication of the number of eggs in the environment.

The World Health Organisation considers that it has become a serious problem when more than 1 percent of the human population is infected. The parasite becomes dangerous when eggs from human waste infect another human and a large larval cyst starts to grow. In humans this larval stage could develop in the brain, damaging brain tissue, causing headaches and epilepsy. In Bolivia, for example, many cases that have been diagnosed as epilepsy are, in fact, neurocisticercosis. There is no cure once the cysts have been formed.

Increased general awareness about the problem may well be one of the keys to this issue, as well as measures on the part of the municipalities. Non-governmental organisations, health workers, schoolteachers and extensionists should all be aware of the problem and the possibilities for preventing it. Preparing pig meat by deep frying or cooking diminishes the risk of transmitting the parasite. Latrines can help to prevent the mixing of human excrement with pig feed.

Clearly, a participatory and interdisciplinary approach is needed, including efforts from the medical and veterinary sciences, as well as from municipalities and producer groups. Top-down approaches, which view extensive pig production as backwards and aim to replace it with intensive pig production systems, do not understand the origin of this zoonosis, which is poverty in the marginal rural and urban areas.

Newcastle disease

Newcastle disease (NCD) occurs throughout the world and is a major cause of mortality in traditional, extensive and intensive chicken production systems. During an epidemic, the disease can cause mortality rates of between 65 and 100 percent. After an outbreak the disease usually disappears for several years, but returns to cause high mortality once all the birds have become susceptible again. The insecurity of occasional sudden high mortalities often discourages families from keeping poultry.

NCD is generally introduced via infected chickens bought at markets. The disease then spreads rapidly through contact among the chickens, or between humans and chickens. The higher the number of chickens, and the closer the households, the more rapidly the disease spreads. The symptoms are sudden death, respiratory difficulties and nervous symptoms in birds that have survived the first part of the disease.

Poultry farmers try to control the disease with all kinds of traditional products based on plant roots, leaves and barks, as well as with human medicines, including aspirin. In Latin America, therapies and prevention strategies are often based on lemon juice, but indigenous technical medicines are often not completely effective in combating infectious diseases. Commercial vaccines that are excellent ways of preventing NCD are available, but these are developed for mainly indoor use in large-scale commercial enterprises.

Vaccination of small numbers of scavenging chickens is hampered by the fact that each animal has to be caught individually to receive an eye drop of vaccine. Moreover, the vaccine has to be used within a few hours of resuspension and has to be kept at a low temperature. NCD-prevention programmes can be successful but they require that groups of chicken keepers be organised. Typically, this is a woman's activity which can also serve other development activities. It is imperative that vaccines are developed to suit the small farmer. Thermo-stable vaccines have been developed, but these still require that the chickens be handled individually, a problem with scavenging birds.

Animal Welfare

Animal welfare seems to be a concern for wealthy countries only. However, it is not just a luxury since the well-being of animals is reflected in their output. Good health, good feeding, good housing and an absence of stress help to maintain the output, but there are trade-offs. For example, animals may feel more at ease when they are left out to pasture or scavenge, but these conditions imply a greater system stress in terms of parasite load, damage to gardens, nitrogen emissions, etc.

Protocols and methodologies are now being developed in Western countries to find a balance between these positive and negative aspects. It is also becoming increasingly clear that high output is not strictly correlated to animal well-being: in many countries it is forbidden to keep animals in cages or to "maim" them unnecessarily, such as by debeaking or dehorning.

Animal welfare is not exclusively a Western issue. In developing countries, particularly in India, there is a strong lobby for the well-being of animals. In cities there are farms where old cows are kept until they die, and activists work hard to improve the lot of animals, for example, by introducing improved harnesses for the bullocks, horses or donkeys that pull carts. Other typical measures include the protection of animals' hooves against hard road surfaces and the introduction of regular watering and feeding and the avoidance of heat stress.

Other, perhaps unlikely, examples of non-Western concern for animal welfare concerns include:

— Fulani pastoralists from West Africa have a very caring attitude towards their cattle. As well as being economically indispensable, the animals are also considered to be members of the family. For instance, they are given names, and "losing an animal feels like losing a child".

— In Papua New Guinea women have been known to cut off a finger joint in mourning for a pig, a practice that is usually reserved only for the death of a child.

Much can be done to increase awareness of these issues. In both Western and tropical societies, a growing category of consumers and producers question the validity of intensive animal production systems. Apart from leading to such problems as diseases and pollution, which are inherent to specialised and intensive farming, these systems provoke resistance because of their industrial character. Indeed, animal welfare has been neglected too long, and products produced in environmental and animal-friendly ways are becoming increasingly popular, even though more expensive. They generate opportunities for urban and rural producers, in developed as well as developing countries.

Livestock Species for Urban Conditions

Goats, sheep, cows, horses, camels, chickens, buffaloes, pigeons and many other types of animal can be found in cities around the world. Each of these animals has its own specific advantages and disadvantages. Small animals, for example, are particularly adaptable to backyard conditions, they require little start-up capital, it is easy to sell them and they reproduce quickly. Aquaculture represents an interesting diversification of agriculture at the periphery of cities. In Abidjan (Côte d'Ivoire), fish are fed rice bran, and slaughterhouse residues and manure are used as fertilisers to produce feed. Tilapia culture in Southeast Asia is currently spreading and intensifying. In Thailand, peri-urban enterprises use processing wastes and other inputs from cities.

Pig Farming

Pig farming is common in the urban areas of many countries, other than those in which the Islamic or Jewish religion is prominent. Pig keeping adapts well to the family level where the role of women is very important, both in collecting household waste and in looking after the animals. Pig production implies a significant reuse of household waste as feed, but the waste of commercial enterprises (bakeries and vegetable and fruit markets) and industrial activities (breweries and abattoirs) is also useful.

Pig farming allows households to generate supplementary income in peri-urban squatter settlements (slums) in, for instance, Montevideo (Uruguay) and Port-au-Prince (Haiti). In these areas the activity is generally linked to the widespread practice of collecting, sorting and selling household waste to the local recycling industry.

Most pig breeders are small producers who have one or two sows and raise their animals from birth to fattening prior to slaughtering. Usually they sell suckling pigs (either slaughtered or alive) to intermediaries and slaughterhouses or directly to consumers. Other producers are more specialised: some raise the animals from birth until they are weaned, when they are sold to fatteners; others sell slaughtered or suckling pigs. Still others fatten the animals until they weigh between 90 and 120 kg, when they are sold or slaughtered.

In large-scale commercial units it is common to keep pigs until they reach a specific standard weight; subsistence and semi-commercial farmers tend to slaughter their pigs when they need money. Typical problems associated with pig keeping are caused by fears that pigs spread disease, that young piglets in particular will be involved in car accidents and that pigs cause noise and public nuisance.

Ways to cope with these issues include providing housing, reducing the number of pigs so that they can survive and grow on local waste and keeping them in a hygienic way. Cross-breeding of pigs is common but tends to be useful only to commercial growers since the resulting animals generally have a different body composition and flavour than those that are preferred locally.

Poultry Production

Poultry production can be divided into traditional backyard, semi-commercial, commercial and industrial poultry systems. Large poultry enterprises are found every where, but small-scale poultry production is also widespread in urban areas. Poultry includes chickens, ducks, turkeys, etc., but this section refers only to chickens.

Chickens are kept for many reasons, including for gifts and ceremonial activities. A major purpose of chicken keeping is to supplement household revenues in terms of food and cash. When daily wages are low the sale of even only a few eggs can be a very substantial contribution to the family income. Urban households with relatively high incomes also keep chickens, especially laying hens, because they believe that eggs produced at home are of higher quality than those found at the market.

The special liking that people have for village-raised chickens (more taste, tougher meat) is reflected in a higher price for these animals. Specialised, large-scale and market-oriented chicken production is found in and around urban areas across the world, wherever access to young chicks for broiler and egg production, inputs (feed) and markets is relatively easy.

This section highlights two important modes of urban poultry:

— traditional family poultry with scavenging chickens;

— backyard chicken production.

The two systems are fundamentally different from each other. Farmers with chickens in a scavenging system live under different economic and social conditions than those who keep backyard, often enclosed, animals. The latter produce mainly for the market and are therefore ready to spend cash on concentrates, disease prevention and housing. The former, in the first place, keep animals for home consumption or sale when need arises, and they do not invest because they want to keep their way of chicken keeping as economic as possible.

Unfortunately, extension workers often mistakenly believe that, with education and some inputs, a scavenging system can be transformed fairly easily into a more intensive backyard system. This is a typical case in which an apparent problem in livestock production can be overcome by reorienting extension workers: farmers are not necessarily mistaken in adopting one mode of production in preference to another and they tend to behave according to rules based on tradition and economics.

Scavenging birds

A flock of scavenging birds usually consists of some adult chickens, one cock and some pullets. In several areas, such flocks also include hybrid layers. Housing is absent or very rudimentary: often an improvised pen of old boxes turned upside down, baskets, etc. Grains, legumes and food wastes are given via provisional feeders or thrown on to the soil. Water is supplied only in the dry season. Nests are made of leaves or old clothes to keep costs low, and scavenging chickens eat mostly household food leftovers, second-quality grains and milling by-products.

Scavenging chickens need to be protected from predators during the night. In many cases the chickens spend the night in trees, which can be covered by a metal guard. Farmers in Africa often use a small shed on poles near their homesteads. Dogs sleep nearby and during the day cocks protect the hens from, and warn them of any danger from predators.

Low-input (scavenging) poultry keeping is generally regarded as a women's activity. Husbands often become interested when more cash is involved. Government officers, who have access to cash and credit, are particularly likely to take up the activity as a second source of income, but chicken production is highly preferred by women and can be an important tool to increase their income. Small informal groups to help women start this activity have been shown to have significant impacts.

Backyard system in pens

This system refers to small-scale activities with confined chickens that are undertaken by families in their backyards. Such activities demand care and the input of concentrates, whereas scavenging chickens balance their own feed intake and need fewer inputs. Young broiler chicks are usually bought when they are one day old and fattened over a period of six to eight weeks.

Governments and non-governmental organisations (NGOs) have been instrumental in initiating such activities. Help with vaccination schemes, input supply and marketing can be particularly useful during the difficult first few years. Feed can account for up to 70 percent of total production costs, so it is important that it be used efficiently. It is worth investing in good feeders that keep losses from spoilage down to a minimum. These backyard activities are popular because of their quick cash returns resulting from the short production cycle.

Layers for egg production are kept in an intensive way and the investments are longer-term than they are for broilers. Housing for layers is more important and, if the animals are enclosed, a balanced ration is indispensable in preventing nutritional stress. Electricity is sometimes used to provide the longer daylight hours that are necessary to induce egg production.

For layers, either one-day chicks are raised until they reach the point of lay, or farmers purchase pullets. The first method is more common but has disadvantages compared with the second one. Young chicks are non-productive for a long period and require special conditions with respect to housing and management (vaccinations). Chicks almost always need heating for the first few days of life, and this can be provided, for instance, by a simple brooder with a paraffin lamp. They are usually vaccinated against several diseases, of which Newcastle disease and fowl pox are the most harmful and common. Vaccination schemes vary according to local conditions.

The production of young chicks for intensive poultry systems is a specialised job that requires parent stocks and a good hatchery. Where conditions for broiler and egg production are in place (concentrates, markets, vaccines), day-old chicks are always available. Small-scale hatcheries can be found on the market, but the results of local hatching are often disappointing (low hatchability and vitality). The cost prices of chicks will therefore be higher, and these kinds of hatcheries are more attractive to the breeders of exotic breeds, e.g. fighting cocks.

In the urban backyard, an old building, or even a room, is often used as a poultry shed. For confined chickens a good pen with easy waste disposal is required, for example, a layer pen that is 1 m high and has a floor of wire mesh or wooden slats. Pens are normally cleaned when the litter has absorbed manure and between batches of animals. If cleaning is complicated, farmers will not do it often enough and health and environmental problems are likely to arise.

Manure is a valuable fertiliser for gardens, so investment in a manure collection system that is easy to operate is therefore repaid by higher crop production, as well as increased feed for livestock. With regard to the latter, proper bedding makes the dung even more palatable and such bedding can also be used for dung ensilage. The use of hormones or other medicines makes dung less suitable for cows (and the meat for humans).

Disease pressure in an enclosed environment is different from that in open, scavenging systems. Coccidiosis occurs particularly frequently when animals are confined, but timely application of coccidiostatica is effective. Worms, bronchial viral infections and bacterial diseases such as coli infections and cholera are frequent. Treatment and vaccinations can solve those problems in a technical sense but they need to be suited to the local conditions and perception to be successful.

Rabbit Keeping

Keeping rabbits in urban areas is common in many countries, including Indonesia, Mexico, Ghana and Egypt.

In some cases, rabbits provide an essential source of high-quality food (protein) for the family, in other cases they provide income or "pet" value for children. Rabbit keeping is also a typical emergency activity. In towns where urban rabbit raising is common the animals are kept in cages on rooftops, in gardens and even in empty rooms. Units of between four and six does and a buck are frequently observed.

The people who live in towns are generally more wealthy and can sometimes afford to buy metal cages, although these are not strictly necessary; any simple wooden or bamboo cage may do, as long as strict hygiene is maintained. Generally, people start with a pair of rabbits, but it is advisable to keep four to six does, if enough feed is available. It is also possible to maintain only one doe and mate her with a neighbour's buck.

Common meat breeds have an adult live weight of about 4 kg, and can be slaughtered when they weigh between 2.3 and 2.4 kg (live weight). It is not necessary to keep only pure breeds and genetic traits in the system described have been exploited only partially. Four parturitions per doe per year may be regarded as a reasonable standard. Under most circumstances, it is possible for an average of five rabbits per litter to arrive at slaughter weight. A total production of about 20 animals per doe per year is very satisfactory, and any number over ten is fair. A unit of five does can produce more than 50 rabbits a year, meaning 2.3 to 2.4 kg of rabbit per week. This level of production can be as much as doubled, especially if concentrates (bran) are fed to the does during late pregnancy and at the start of lactation.

Breeding animals should be replaced continuously to avoid decreases in production. As a general rule, one out of ten adult does should be replaced monthly, but low-yielding does should be eliminated immediately. Some animals continue to produce for several months in the second or even third year of production. Bucks are replaced annually and start mating at the age of six months. The exchange of males to reduce inbreeding is not as important as is often believed, because rabbits are territorial animals and inbreeding is common, even under natural circumstances.

Up to a certain degree, reduced fertility is even an advantage. Rations, consisting mainly of kitchen and vegetable waste, are often poor and unbalanced, with protein and energy contents that are too low to allow does to produce enough milk for a litter of more than five or six puppies. Larger litters are less likely to survive, and it is more economic to wean five animals from a litter of five than one or two from a litter of eight or more. For the same reason, does are not usually mated while they are still lactating, even though nature allows them to be mated on the day of giving birth.

If good-quality feedstuffs are not available ad libitum, it is better to mate only after the previous litter has been weaned or even to wait for some weeks until the doe has gained weight. The day of mating must be registered—record keeping is a useful practice with rabbits. After two weeks the doe has to be checked for pregnancy and, if negative, should be mated again. The palpation technique for discovering pregnancy needs practice. Twenty-eight days after mating, the mother will prepare the nest, ideally in a specially provided box, and clean straw should be provided. The number of successfully weaned puppies should be the main selection criterion for does. The young does and bucks to be kept for reproduction must be chosen from the litters of does that have shown to be good mothers over at least three parturitions.

Rabbits can be fed with vegetable wastes or (dry) grass that is collected daily: a certain quantity should not be eaten immediately. This is a very important point and if there is not enough feed, it is necessary to reduce the number of animals. Clean water must always be available, although rabbits survive without drinking if enough fresh grass or other vegetal resources are at their disposal. In hot climates feeding should take place in the late afternoon so that rabbits can eat during the coolest period of the day. To avoid diseases, the cage floor must be cleaned daily.

Keeping of Guinea Pigs

The keeping of guinea pigs is somewhat similar to that of rabbits. It can be done in urban and rural areas and guinea pigs alleviate food deficiencies in places where other kinds of animal production are difficult. As do rabbits, they eat any kind of grass or leaves, and a small daily quantity of fresh feed per animal is sufficient. Guinea pigs need very little space; a fenced area of about 1 m^2 is sufficient for eight to ten does and one buck. Any material, from a cardboard box to bricks, can be used.

Management is very simple because there is no need to interfere in mating or nest preparation. After a gestation period of approximately nine weeks, an average of 2.5 animals per litter are born. Newborns are able to feed by themselves immediately and can be weaned after two weeks. Each doe can produce about eight to ten animals per year, which means about 100 offspring from a group of ten does, corresponding to nearly 1 kg live weight per week. Although a heavier breed occurs in Peru, the smaller variety is more common and can be found all over the world. The adult live weight is 800 to 1 000 g and animals should be slaughtered when they have reached about 450 g.

Straw, dry leaves and even newspapers can be used to absorb the urine and must be changed when they become wet. Animals must be provided with water and fresh grass, leaves or kitchen wastes every day, and care should be taken not to throw the feed on to the litter—more than 15 percent is spoiled in this way. A small rack is sufficient to avoid feed spoilage and to improve hygiene.

Disease incidence and mortality are very low but, at the first suspicion of sickness, as with rabbits, the animals must be slaughtered and can be eaten if they are large enough. Raising rabbits, particularly in towns, can be difficult because of lack of space, insufficient feed availability, insufficient resources for buying or constructing cages and nests and the complexity of managing the animals' reproduction efficiently. In these respects, guinea pigs may be a suitable alternative because they are smaller and easier to keep.

Pigeon Keeping

Pigeon keeping is very popular in the Mediterranean region; for instance, in the Nile delta, dovecotes are common in both rural and urban areas. Pigeons can contribute substantially to household diets and income. They do not compete with other animals for space and feed; if fed by their owners, the birds tend to remain in the neighbourhood, but they are able to find feed within a radius of 15 km, thereby making use of the different vegetation cycles of local plants. In low-input systems, feeding is necessary only during the short period when the animals are getting accustomed to their new home.

Pigeons adapt easily to urban conditions and are a common sight as they scavenge in town squares and markets. Pigeons nest in natural or artificial holes in rocks, caves, towers and other buildings. Dovecotes are normally located on rooftops, which makes pigeon keeping possible even for people who live in multi-storey buildings—an important consideration in developing country towns. Dovecotes are usually constructed of wood or from mats attached to a wooden frame. Pigeons reach maturity at about six months of age; incubation lasts 18 to 20 days and both sexes are involved in the hatching.

The newly hatched squabs are fed a special substance that is produced in the crop of the adults ("crop milk") and, later, eat regurgitated feed. Each clutch contains two eggs and, if no supplementary feed is given, pigeons brood about five times a year. Ten couples can produce eight squabs per month, which can be collected at the age of about four weeks, when their live weight will have reached about 400 g, just before they are able to fly. In more intensive systems, females lay two new eggs when the squabs are still present in the previous nest, requiring a two-nest structure. Surprisingly, in spite of being easy to raise and cheap to produce, this species is very rarely considered in urban food security programmes.

Stakeholders in Urban Livestock Keeping

Among the stakeholders in urban livestock systems are the shantytown women and children who make a living by, among other activities, feeding their food wastes to rabbits, guinea pigs or poultry. Other stakeholders are government officials, entrepreneurs, shopkeepers, leaders of dairy cooperatives, feed and medicine merchants, retailers and wealthy consumers.

The differences among stakeholders imply different interests and perceptions regarding urban livestock. Government officials may know little or nothing about the socio-economic and technical aspects of animal production but they can be zealous in protecting their own private interests. Entrepreneurs may be retired or active government employees, their wives, single women or unemployed casual labourers; they may be schoolchildren or industrialists who know how to make good use of government subsidies and tax laws. Combined with the different forms of urban livestock, the result is a wide range of different ways of perceiving problems and opportunities in urban livestock; and some form of categorisation is required to cope with this variation.

For instance, all those people who own a bullock cart could be regarded as representing one group; other groups could be all those who own two, or three or more dairy cows. Such people depend on their livestock for at least part of their and their families' livelihoods, and they will be reluctant to give up this source. The same is true of single parents who are raising extra income, improving their children's diets or earning money for school expenses by keeping animals, which also help to dispose of kitchen waste or by-products from the local industry. However, not everybody within the same level of urban society shares the same positive view of urban livestock keeping. For example, a livestock keeper may live next door to the mother of a child who comes home with clothes dirtied by animal excreta, or there may be a neighbour who gets angry about goats causing havoc in his vegetable plot.

A second category of people consists of local legislators and civil servants. Their main concerns are about disputes in their neighbourhoods, or about raising some extra money by fining people for breaking the law. Many administrators regard urban livestock keeping as a sign of backwardness, but there are exceptions. For example, much of the urban dairy business of Dar-es-

Salaam in the United Republic of Tanzania is carried out by the city's civil servants as a sideline.

A third category of stakeholders in urban livestock are those administrators, and often academics, who are professionally geared to see only one aspect of what ever it is they are looking at. For example, people who work in the public health office will be concerned only about public health; tax collectors will worry only about tax (and fee) collecting; those in charge of urban housing will concentrate on construction plans and ignore the animals; and veterinarians will be interested in zoonoses and private practice opportunities. This description may be somewhat exaggerated but it does characterise the ways in which many employees of government offices tend to function.

A health official cannot compromise on public health even when animals represent an essential form of income that helps local people to survive; a tax-collector cannot tolerate informal arrangements that are difficult to administer, even when they form an essential part of local livelihoods. Reports from such sector departments and/or discipline-oriented professionals are very likely to emphasise problems in whatever area they are regulating or studying. The identification of creative and holistic solutions is not part of their mandate, nor have they been trained to do so. One possible way of improving livestock systems, therefore, lies in the (re-)training of government officials to use their rules to help strengthen the positive aspects of urban livestock while overcoming the negative ones.

A fourth category of stakeholders in urban agriculture is that of the national and international planners and policy-makers who are concerned with the production of sufficient food, of plant and animal origin, to ensure that people get enough to eat. These people tend to emphasise the fact that urban livestock and agriculture produce only a fraction of the dietary food requirements for an urban population, and they rightly concede that urban livestock and crops can never produce enough. A combination of the possible pollution and health risks of keeping animals in close proximity to people and the competition with other sectors, leads planners and policy- makers to suggest that the production of food from animal origin is better shifted to larger specialised units outside the city. As with the discipline- and sector-oriented people, these people also tend to overlook the fact that urban livestock can fulfil many roles besides producing milk or eggs, and that a small contribution to food security can make a lot of difference at the family level. Among planners and policy-makers it would be beneficial to increase awareness about the essential roles that animals can play in the economy, sociology and ecology of cities.

A fifth but very relevant category cuts across all the others. This categorisation distinguishes between, on one hand, so-called "linear" thinkers who are interested in only one aspect, e.g. food production and, on the other hand, "non-linear" thinkers who consider several aspects and interests to reach holistic approaches to the issue. The two lines of thought should supplement each other, but in practice they tend to run counter to each other. Successful action is often the result of local creative initiative rather than general across-the-board legislation. The distinction between linear and non-linear is real but vague; in this publication such a distinction facilitates an understanding of the technological issues that exist in the complex systems involved in urban agriculture. In essence, these issues arise in any complex system, whether urban livestock, mixed farming or manufacture/ supply chains. The unique aspect of urban livestock systems is that they present a microcosm of livestock systems in general, by relating to pigs, cows, draught animals, pets, milk, social and biophysical functions, etc. Because urban livestock systems include the direct involvement of such a wide variety of stakeholders, they provide a unique example of working with complexity and a good opportunity to learn about similar aspects of agriculture in general.

Factors Favouring Urban Livestock Production

Many modern urban livestock systems occur in tropical regions, but their existence has been described for a long time, for example by the agricultural economist Von Thünen, who researched German conditions some 150 years ago. In his "location theory" he showed how intensive livestock systems tend to occur right in and around urban centres, together with other intensive forms of agriculture such as vegetable production.

The production of forests for energy and timber took place at a somewhat greater distance from the towns, followed by cropping and more extensive forms of agriculture and, ultimately, by extensive livestock production. Obviously, the location theory should not be taken too literally; fossil fuel has replaced timber and the general pattern may be disturbed by rivers, roads, mountain-ridges, etc. Still, the basic outline of Von Thünen's theory should be kept in mind when discussing the emergence of urban livestock systems in any part of the world.

It can be hypothesised that (peri-)urban agriculture and urban livestock systems originated from several complementary factors, including:

— the availability of a choice of high-quality feeds such as by-products of the food processing industry (bran, oilseed cakes), hotel refuse and kitchen waste. These feeds may be available in a scattered fashion in the countryside, but in the city they can profitably be combined into nutritionally and economically more optimal ratios;

— a strong local demand for food and power from animals, often resulting from the presence of wealthy consumers. Note that a moderately wealthy population tends to consist of a poor and a rich sector. The rich, in particular, verge towards increased consumption of food from animal origin;

— the potential of vegetable and livestock units to provide a high turnover, which gives a relatively high income on small plots of expensive land;

— the presence of significant proportions of poor people and wealthy investors, both of whom find an extra source of food security and/or income through the keeping of animals.

Other factors that affect the tendency for urban livestock systems to emerge include the availability of specialised services such as veterinarians and processing industries and the need for short distances between producers and consumers for commodities such as fresh milk and "hot" meat. These factors combine with the availability of more land at a greater distance from the city, where staple crops such as cereals can be grown together with feed and where animals can be raised for fattening in the city or to serve as pets for urbanites. The upland town of Batu used to have a thriving market of pet rabbits, which were sold at weekends to the wealthy upper class from nearby Surabaya and Malang.

Similarly, cattle and sheep fattened around London were raised and grown in Scotland; animals raised in rangeland areas of the southern Sahel are finished in cities such as Abidjan; and urban dairies of Karachi and

Mumbai transport their lactating animals for long distances.

Problems in Urban Livestock Systems

Many urban livestock systems are undergoing a period of crisis. In many government offices, there is a feeling that urban livestock systems are under pressure from a variety of issues. Many of these issues are clearly problematic, others are open to several interpretations.

Urban livestock systems as a potential hazard to public health. Potential hazards for public health range from poor hygiene, caused by the presence of dung, flies, parasites and zoonoses, to the difficulty of controlling product quality when food from animal origin is used directly by consumers. For example, livestock keeping without a proper sewage system may favour mosquitoes that transmit malaria and some important viral diseases, such as yellow fever, dengue, Rift Valley fever and West Nile virus.

Zoonoses are diseases that affect both humans and animals. They are more likely to spread when hygienic conditions are poor - for example, commercial forms of livestock keeping in urban areas are particularly favourable to the multiplication of rodents because these systems require the storage of animal foodstuffs. Rats are the reservoir of, for instance, hanta virus, an important human disease in Asia.

In Africa and Latin America a human tapeworm makes use of scavenging pigs to complete its lifecycle. Pigs feeding on human excrement containing tapeworm eggs become infected by cysticerci (the larval stage),while humans who have eaten insufficiently heated pork meat become hosts to the adult tapeworm. The disease becomes a serious problem when other humans, such as playing children, are infected with tapeworm eggs from human excrements. The housing of animals, maintenance of hygiene and proper storage of animal feeds are typical responses to these problems.

Urban livestock systems as a potential hazard to animal health. Poverty, ignorance and a lack of veterinary services favour the spread of human and animal diseases. In addition, where there are poverty and administrative incompetence, these diseases are less likely to be noticed by veterinary officials. In Ghana, the first outbreak of African swine fever (ASF), an endemic disease in East Africa, occurred among backyard animals.

The disease had time to become rooted in backyard pigs across Ghana before the cause and significance of the subsequent mortality were discovered by the veterinary services. Uncontrolled slaughter aggravated the spread of ASF. In a similar way, it may have been refugees from Mozambique who brought Newcastle disease to Zimbabwe some years ago. The severe mortality in chickens all over the country was difficult to constrain, as adequate vaccination campaigns in backyard poultry were complicated to organise.

However, in spite of the thinking prevalent in many government offices, it is incorrect to blame these problems solely on backyard farming. Large losses have been incurred in so-called modern commercial and industrial systems, such as the outbreaks of classical swine fever and bovine spongiform encephalopathy (BSE or mad cow disease) in Western Europe.

Pollution from manure effluent and abattoir waste. Pollution from animal production is caused by a range of issues, some of which are poorly specified and based on conjecture or local observations. Potential problems with dung and urine disposal are obvious; flies breeding on animal and vegetable waste may transmit food-borne diseases, such as diarrhoea. Problems arising from wastewater and rejected meat from abattoirs (slaughterhouses) are less obvious to most of the general public. When the building construction is adequate, there is proper meat inspection and strict removal and destruction of rejected carcasses, abattoirs do not pose a risk for human health.

The pollution issue needs to be balanced with the fact that livestock keeping using the resources available in (peri-)urban areas contributes substantially to the cleaning of local waste. Chickens keep the backyard clean and create value from leftover food, while pigs utilise hotel waste and by-products from agro-industries, thereby reducing the amount of organic waste to be disposed of. Some cities actively promote programmes to recycle poultry waste for agricultural purposes (Ghana). Urban dairy animals in Mexico City clean leftover food (tortillas) while their dung is circulated to urban gardens.

Urban livestock systems are often old-fashioned, unproductive and inefficient. The inefficiency of urban livestock systems is a matter of perception rather than objective fact. Remarkable changes of attitude regarding urban livestock have been achieved by special teaching programmes, for example in Mexico, resulting in urban youth and veterinary graduates understanding the value of livestock keeping in and around cities.

Typically, non-governmental organisations (NGOs) and urban planners are far more positive than government officials who are trained to see and control only one aspect of the whole issue - health or product quality problems, for example - without taking all the other factors into account. NGOs and local stakeholders tend to regard the sector from the point of view of its profitability and usefulness for the producer and local consumers, whereas much academic training tends to emphasise the biological performance of animals, often based on standards and regulations from countries that are quite inappropriate to local conditions; (re-)training and institutional change are only two of the important measures required to improve urban livestock keeping. Definitions of productivity and efficiency refer to a ratio of output over input, not to high production levels as such.

At the same time, low output does not necessarily reflect low productivity. For example, scavenging chickens producing only one or two eggs per week require hardly any inputs and can be economically profitable in terms of cash return per cash input. The equation of productivity with efficiency can also be misleading because it does not take into account the multiple goals of much, particularly small-scale, urban livestock keeping. For example, pigs clear away waste as well as producing meat, while guinea pigs, rabbits or even goats provide social satisfaction as pets along with opportunities for producing milk and side income.

Only a small proportion of urban food requirements comes from urban livestock. A major concern regarding urban livestock systems is that the (growing) demand for food, particularly for wealthier consumers, from animal origin in urban areas is outpacing the production potential of urban systems. Urban livestock may never be capable of producing enough food for the entire city, but the urban livestock system's share of food production can be substantial for some sectors.

An important point for understanding urban livestock systems is, however, that animals have many functions. In addition to providing food, they:

— provide social comfort and emotional attachment;
— convert waste;
— can be a source of income;
— serve as a savings account;
— provide energy (draught power).

Indeed, small-scale enterprises keeping rabbits, guinea pigs or even poultry, goats or cows often start up in areas where no similar activities existed previously and generally do not demand public money. For example, rabbit keeping was adopted in London during the Second World War as a solution to issues of food scarcity. The smaller forms of animal production are very dynamic and adaptive. They provide income and useful products to the city's inhabitants and they contribute, even as a side activity, substantially to the livelihood of many households. Generally, even in cities, a rural landscape is appreciated; this is reflected, for instance, by the higher prices of houses facing farm plots or parks instead of apartment buildings.

Variation in Urban Livestock Systems

Urban livestock systems face problems but they also offer opportunities. The opportunities depend on people's different perceptions of reality and their, often unexpected, creativity and survival skills. One way to overcome the problems associated with urban livestock is to concentrate on the positive aspects; the (re)training of policy-makers is important in this respect and the redesigning of curricula may be required.

The problems of complex systems such as livestock keeping in urban areas can be dealt with in several ways; they have, in the terminology of modern system theory, several coping strategies. For example:

— The smell from urban poultry can not only be overcome by killing or moving the chickens, but also by building other housing systems or by improved cleaning of sheds.

— Animal disease in urban livestock systems can be addressed by seeing it as a reason to do away with the system altogether. It can also be treated as a management issue related to socio-economic conditions in the surrounding slum or suburb.

— A community can embark on a health education programme that would automatically take care of the animal disease rather than address symptoms only.

— Re-allocation of animals would be one way of solving problems related to animal excreta. However, an urban gardening or biogas production programme could be a better solution by converting the excreta into a valuable resource rather than waste, thus maintaining a "vibrance" and social activity rather than "killing" local initiative.

— Improved processing of animal produce is one way of avoiding wastage, but quicker retail and better hygiene for local consumption might also be an option. In many instances, sophisticated slaughterhouses and other processing facilities result in higher prices and, ultimately, a continuation of illegal practices while publicly funded processing plants remain idle.

— Perceived problems can be solved to some extent by balancing education about problems with education about opportunities rather than by ridding the com munity of their livelihood and local creativity.

6

Environmental Impacts of Livestock Production

The livestock production systems have direct and indirect effects on the environment. The livestock-environment interactions (LEI) can be negative, but also positive interactions occur. The LEI are related to the stage of the production process:

(a) the production and delivery of inputs to LLR systems;

(b) directly to the production process itself; and

(c) the processing and marketing of the products from LLR systems.

External Inputs to LLR Systems

The main external inputs of the LLR production system are: (1) feed, mainly in the form of concentrates; and (2) livestock for further fattening (or milking).

Feed Production

The effect on the environment of the demand for concentrates has been described extensively by Héndy et al. The main aspects are:

— The requirement of land for production of concentrates. The environmental impact includes competition with humans for food, changes in land-use, increased land pressure, effects on soil, water and the atmosphere (use of fertilisers and herbicides/pesticides), and effects on soil erosion as a result of intensified cropping.

— The utilisation of crop-residues and by-products of the agro-processing industry. Large quantities of these types of products are consumed in LLR systems and converted into useful high quality food. If not used, the residues and by-products would form a gigantic environmental problem.

— The production of forage for LLR systems. The intensive forage production on a small land area can absorb part of the manure produced by the animals and reduces the land area required for animal production making more land available for wildlife, forest, rangeland etc..

— Energy requirement for transport of feed to feedmills, the feed milling and mixing process, and the transport to the farm. Transport and processing requires fossil energy resulting in emissions to the air/atmosphere of CO_2.

Livestock for further Fattening and Milking

Cattle and lambs are required for fattening in feedlots, and dairy cows and she- buffalos are required for the urban dairies. The main environmental aspects are:

— Cattle and lambs for fattening. Increased numbers of livestock required for fattening put extra pressure on rangeland and/or results in expanded rangeland areas at the expense of wildlife and nature conservation. On the other hand, the lamb fattening system in the Middle East has been developed and is being subsidised to reduce pressure on the rangeland and prevent rangeland degradation.

— Dairy cows. For the urban dairy systems dairy cows come from the rural mixed farming areas. This trade itself does not have a direct environmental effect on the mixed farming system but because the best animals are selected it forms a drain on the genetic resource.

LEI-related to Production Process

LEI are related to (1) the production of manure; (2) the use of drugs, growth stimulators etc.; and (3) the use of fossil energy.

Manure Production

The main environmental effects are caused by emissions from manure in the stables, during storage, after application to the land or when manure is simply dumped. Emissions are in the form of nitrogen, phosphorus, methane, organic matter and possibly heavy metals. Manure is also a useful fertiliser and an input for crop and pasture production. Biogas production is another form of manure utilisation. Brandjes et al. deal extensively with most of these aspects. The main issues are:

— *Manure as fertiliser:* The use of manure can improve soil fertility. In certain areas manure is a valuable commodity and sold especially for vegetable production. The small areas of land required for intensive roughage production can absorb a part of the manure from the LLR system.

— *Manure for biogas:* Manure can be used for the production of energy (Methane) and the remaining liquid slurry can be used as fertiliser. The manure from intensive systems is not very suitable for use as fuel in a dried form.

— Nitrogen (N) emissions are in the form of (i) volatilisation of NH_3; (ii) volatilisation of N_2 and N_xO_x; and (iii) run-off and leaching of N-compounds.

(i) Volatilisation of NH_3 during storage and application on the field. NH_3 emissions cause acid rain and eutrophication of the ecosystem.

(ii) Volatilisation of N_2 and N_xO_x occurs in anaerobic situations (in lagoons and after application to the soil as (by)-products of nitrification and denitrification processes. N_2O is especially harmful as it contributes to global warming and breakdown of the ozone layer.

(iii) Run-off and leaching of N-compounds (nitrate etc.) during storage and after application to the land. These compounds can reach the ground water and make the water unsuitable for drinking water, eventually contribute to eutrophication of the surface water.

— Main P emissions are through run-off and causes eutrophication of the surface water. However, if P fertilisation is in excess of crop P requirements for longer periods, P saturisation of the soil occurs leading to P leaching to the ground water.

— *Odours:* animal manure contains a number of volatile organic compounds with an obnoxious odour The compounds do not have a direct negative impact on the environment except that they are a nuisance to the people of the surrounding area;

— Methane (CH_4) emission is from two sources in LLR systems (1) direct from the digestive process in the rumen; and (2) from the anaerobic decomposition process of organic matter in the manure during storage. Methane causes breakdown of the ozone layer. The impact domain study Methane will deal with this aspect of environmental impact;

— Heavy metals in the faeces can form a problem where high levels of manure are used as fertiliser, particularly if removal of heavy metals from the land is low, e.g. in case of predominant livestock farms.

Drugs, Herbicides and Pesticides

The following categories can be distinguished:

— *drug residues*: residues in animal products following preventive or curative treatment of diseases;

— *pesticides residues:* pesticides in products following spraying or dipping for controlling external parasites;

— *growth stimulators*: effects of the use of hormones etc. to stimulate and regulate growth;

— *pesticides/herbicides residues*: herbicides which enter the system through concentrates and crop-residues and appear as residues in the animal products or in the manure.

Fossil Energy Utilisation

Fossil energy is used as energy source for the operation of equipment and transport. Emissions are in the form of

CO_2, SO_2 and N_xO_x contributing to global warming and acid rain problems.

LEI-related to Processing and Marketing of Animal Products

Environmental effects are in the form of waste production from slaughterhouses, tanneries and dairy processing plants and for the use of fossil energy for transport and conservation (e.g. chilling) of products. These aspects are described in the impact domain study 'Environmental effects of animal product processing' by Verheyen et al.

Environmental Effects of Slaughterhouses

Emissions are in the form of:

- *Solid waste:* manure, paunch, hooves, horns and solid slaughter offal and by-products. Most of the solid waste can be composted and used as fertiliser, but may pose a threat to human health and surface water if not treated well.
- *Waste water:* water from cleaning can contain slaughter offal and by-products (e.g. blood). The waste water mainly contains organic matter. The quality is measured in Biological oxygen demand (BOD) i.e. the quantity of oxygen required to break down the organic matter.
- *Volatile compounds:* emissions of volatile compounds is mainly from the use of fossil energy, further from the singeing of pig skins for hair removal, and for the further processing of meat (smoking).

Environmental Effects of the Tanneries

For the tanning of 1 ton of raw hides around 300 kg of salts and minerals are required. Chromium is the main tanning agent and at the same time a major polluting factor as it is highly toxic. Emissions are in the form of:

— *Solid waste:* in the form of scrapings of the raw hide (meat offal etc.), and chromium containing scrapings and cuttings of semi- or fully tanned hides. Discarded leather products also contain ± 3 % chromium.

— *Waste water:* from cleaning, soaking etc contains organic material, salts and chromium.

— *Volatile compounds:* emitted to the air mainly from the use of fossil energy and the use of dyes for finishing leather products.

Environmental Effects from Dairy Plants

Emissions are in the form of:

— *Waste water:* containing residues of milk and milk products; whey from cheese production is a major water pollutant in some cases.

— *Volatile compounds:* emitted to the air mainly the result of use of fossil energy, and limited dust emissions in milk powder manufacturing.

— *Whey and other dairy by-products:* which can be, and are increasingly, used for feeding calves and pigs.

Following processing, emissions to the air also occurs when energy is used for transport to the retailer and consumer as well as during storage (e.g. chilling). Waste is also produced when meat and milk products go bad and are disposed of.

Magnitude of Interactions

In the description of the magnitude and relevance of LEI the following aspects are important:

— the flow of nutrients;

— eco-toxic effects in the production system;

— effects on limited resources, e.g. soil conservation, water, genetic resources and others; and

— energy requirements.

The major problems that arise from quantifying LEI are the definition of indicators and the availability and reliability of data. Indicators for this particular study have been developed in the impact domain studies, however, due to the variability of the production systems concerned the parameters and indicators cannot always be applied directly. The production systems cover large geographic areas and the statistical basis is not always available and reliable enough to carry out the quantification. Another major problem is the quantification of the indirect LEI of LLR systems.

Nutrient Excretion

In Table 1 estimates are given of the total manure production of the different sub-systems within LLR systems.

Table 1. Estimated annual nutrient excretion in the different LLR-systems

	Total N	*Mineral N*	*Total P*
Feedlot - beef fattening	497	348	133
Veal production	29.4	-	3.6
Sheep fattening	18.7	12.7	4.0
LL in EE and CIS	n.a.	n.a.	n.a.
Urban dairies	n.a.	n.a.	n.a.

The estimations of the manure production from feedlots in the USA and from veal production in the EU are based on reasonably reliable data on animal numbers, feed rations, etc.. Estimates for sheep fattening in WANA are less reliable as only little information is available on feed rations and animal numbers. Lack of in formation on feed composition and, more important, animal numbers in the EE and the CIS and in the urban dairy system precludes a reliable assessment of the manure production.

Manure Management

Solid storage is the most prominent manure storage system. As most feedlots are unpaved, leaching losses may be considerable, depending on the type of soil. Run-off water, containing considerable amounts of dissolved manure particles and nutrients, is increasingly being collected and treated in lagoons, before being discharged to surface waters.

Nevertheless, imperfect lining of the feedlots and overloading still results in part of the run-off being discharged without treatment. This treatment in lagoons does not affect the height of the nutrient losses, but reduces the negative effects of run-off on surface water. Slurry systems are beginning to become more important.

Manure surpluses hardly occur at regional level: even on a county basis the manure production/ land-base ratio is low to medium. However, due to the size of the large-scale feedlots where well over 50,000 head is marketed annually, major manure disposal problems do occur. Such feedlots produce about 1000 tons N and 266 tons P, while crop requirement of e.g. high productive maize silage is less than 29 kg P; thus such feedlots require more than 9,000 ha to maintain P equilibrium fertilisation. The problems of the enormous land requirements are aggravated by:

- Underestimation of actual fertilisation rates: it is not uncommon for some producers to apply 2-5 times more manure than estimated.

— Incorrect habit to base manure application rates on N requirements of crops and N availability in manure, thus risking P over-fertilisation and often not considering N residual effects.

— Lack of manure storage: most crops do not need manure application for a major part of the year, while manure has to be removed after each cycle of fattening.

Furthermore, the often unpaved, open feedlots incur high runoff, volatilisation and, depending on soil type, leaching losses. The practice of runoff being prevented from entering surface water until it has first treated in lagoons is increasing. However, these mainly solve the problem of direct water pollution from organic matter pollution; prevention of eutrophication of surface water by nutrients is not adequate as nutrients, particularly P, still enter the surface water.

Manure from veal production is exclusively stored in slurry systems. As veal manure has a very low DM content of about 2% and a low nutrient content compared to other types of cattle manure, the fertilising value of veal manure is very low. The high water content of veal manure also results in high transportation costs. Consequently, veal manure is increasingly treated as sewage water, also facilitated by the centralised production of the veal.

Feedlots for sheep fattening are unpaved and only the solid manure is collected. Losses are through volatilisation, leaching and runoff, the latter particularly in places with high and heavy rainfall. Most of the remaining solid manure is well used, often sold to farmers in the area. Only in a few cases are feedlots more densely concentrated, particularly near cities, thus posing some manure disposal problems.

Manure management in EE and the CIS is rather unclear. Firstly, several types of storage systems are important, but the proportional importance is far from clear, particularly for something as vague as the LLR systems. Secondly, manure management is changing. Until recently, manure was wasted on a large scale, often applied mainly on the fields nearby the livestock farm buildings, but also directly discharged to surface waters or dumped on wasteland.

Reasons given were the enormous scale of livestock production units, inadequate manure application equipment, cheap fertiliser and/or lack of interest. Underlying reasons are less clear, but partly related to planned economy . However, in various countries this situation is changing considerably. In some countries, large livestock units are being abolished in the process of privatisation, while manure utilisation is improving as artificial fertiliser becomes more expensive or unavailable.

In most countries with urban dairies, manure is a highly valuable product, used as fuel or fertiliser. However, because of the high concentrate rations, the DM content of manure is fairly low, rendering it less suitable for fuel while transportation to vegetable production fields, for instance, is also problematic. Where surpluses are common, manure is often discharged directly into the public sewage system, open surface waters or nearby land.

Heavy Metals

Heavy metals in LLR systems originate from mineral supplements for P and from fertilisers for crop production. An extensive study in the Netherlands on the presence of heavy metals in animal products and manure indicated that with current manure application rates, Cu,

Cd and Zn are the main problem. It is unknown to what extent heavy metals are problematic in LLR systems. It seems reasonable to assume that Cd and Zn concentration in soils will rise to high levels when these soils are overdosed with P. As indicated already by Brandjes et al., heavy metals are unlikely to pose major problems if fertilisation is based on P equilibrium.

Methane Production

The atmospheric concentration of methane (CH_4), currently about 1.7 ppmv (parts per million by volume) is increasing at a rate of about 1% per year and has more than doubled over the past two centuries. Prior to this doubling, the atmospheric concentration of methane remained fairly constant, at least as far back as 160,000 years.

The increased abundance of methane will have an important impact on global climatic change, tropospheric ozone, and the stratospheric ozone layer. Estimates are that methane contributes to about 20% of the expected global warming from the greenhouse effect, second only to carbon dioxide which contributes about 50%. Other related to livestock production estimates, however, show a large variation. There are two main sources of methane emission related to livestock production:

(1) emission from the digestive processes of ruminants; and

(2) emission from the decomposition of animal manure.

Methane Emission from Digestive Process

Methane emissions from the digestive process of ruminants depend largely on the crude fibre percentage of the ration: the higher the crude fibre content, the higher the methane emission percentage of the gross energy intake. However, as lower crude fibre contents are nearly always combined with higher total energy intake, the effect per head is often small.

Different estimates exist for methane emission from beef cattle in the USA. Byers estimates that in the intensive system around 80 kg of methane is produced during the production of an 500 kg animal of 18 months old and the total methane emission from the beef industry in the USA is 2.9 Tg (= million metric tons) per year. Johnson et al. on the other hand, estimate the total methane emission from the USA beef industry as being 3.8 Tg per year based on methane production of different classes of cattle (Table 2). The emission for feedlot cattle in this table is around 39.1 kg per feedlot place, which is much lower than the 65 Kg given by Crutzen et al., while Khalil et al. even assume a production of 103 kg methane per feedlot place.

Table 2. Methane production of different classes of beef cattle 7 Methane production of different classes of beef cattle

Class of livestock	*numbers million*	*Days fed diet per annum*	*% of Methane production*	*Ltr hd^{-1} d^{-1}*	*Tg per year*
beef cows	33.7	365	6.2	262	2.3
calves	38.6	210	6.0	53	0.3
stockers	37.9	150	6.5	202	0.8
feedlot	26.6	140	3.5	153	0.4

Veal calves are exclusively reared on milk and milk replacements, the rumen is not developed and, consequently there is no methane emission from the digestive process.

Little data is available for the methane emission of sheep fattening; all refer to Crutzen et al. who give an annual methane production for sheep in developed countries of 8 kg and 5 kg for sheep in developing countries and Australia. Considering that the sheep in the feedlot systems in WANA are raised on low to medium roughage rations, an estimate of 3 - 5 kg with an average of 4 kg per annum is probably fair. The estimate results in a total methane emission of 10,960 mt from sheep fattened in WANA. The methane emission from LLR systems in EE and the CIS cannot be assessed as information is lacking on the type of ration and average live weight.

Methane Emission from Manure

Methane emission from manure depends on the composition of the manure, the storage, and the distribution system. Potential methane emission is closely related to diet composition, higher digestible rations producing a higher potential methane emission. Under similar conditions, the manure of cattle fed on a high-energy corn-based diet will produce about twice as much methane as the manure of the cattle fed on a roughage diet Thus the reduction of methane emission from the rumen due to higher digestibility is partially countered by the increased methane emission from the manure. Which part of this potential methane emission is actually emitted depends, among other things, on the type of manure storage.

Safley et al., 1992 estimate that the methane emission from beef cattle waste is 1.4 Tg per year of which 0.26 Tg is from beef cattle in feedlots. This assessment, however, was based on 11.2 million feedlot places, thus methane emission from manure of ca. 10 million feedlot places is approximately 0.23 Tg, or 23 kg per feedlot place.

Effect of Landless Ruminants on Global Warming

Though information on the contribution of LLR systems to methane is far from complete and estimates vary widely, partly due to large differences in estimated methane emissions from ruminants, some general conclusions may be drawn. Hopefully the "Methane study" will soon produce more accurate information.

First, methane production per head in LLR systems is higher than from ruminants in land-based systems, partly because of the higher methane emission from the manure and because of higher feed intake. However, methane production per kg output is much lower because of the higher animal production levels in LLR systems. The only exception might be when all manure is treated in lagoons from which process the methane emission from manure is 18 times higher than emitted from range animals.

Second, though LLR-systems are characterised by high fossil energy utilisation, this effect is negligible compared to the effect of the lower methane emission per kg output. Third, manure treatment in lagoons has a large impact on global warming as both methane and N_2O emissions are high, moreover 1 g of N_2O has an effect on global warming equivalent to 15 g of CH_4.

Total Concentrate Consumption

Hendy et al. estimated the total concentrate consumption in LLR systems at 139,443 MT, which is almost 13% of the total world consumption of concentrate. This includes an estimate for consumption of in EE and the CIS of more than 75% of the concentrate consumed in LLR systems. Though we are not in the position to check the values in EE and the CIS, they do not seem abnormal as it is known that high levels of concentrate are common in these countries. Their estimates of concentrate consumption in other countries are higher than expected. This can be due to over estimates of live weights of beef cattle in OECD countries and of fattening periods of sheep in WANA.

Also no distinction was made between basic feeds, specially cultivated for animal production, and by-products, which become available from food processing, industrial use, etc. When waste or by-products, such as inedible grain, soybeancake, molasses and slaughter offal are consumed by livestock, then the burden on the environment from these products will be reduced as part of it will be converted into edible products.

Table 3 gives an estimate of the basic feed and by-product consumption of LLR systems. These estimates are based on average standard feed rations, despite the large differences within each system, often related to differences in feed ingredient prices. Feed requirement estimates for EE and the CIS, and urban dairies are not given due to lack of data. For comparison, Hendy et al. estimated the concentrate utilisation in LLR systems in WANA and OECD countries to be 1.6 and 33 million MT respectively.

Table 3. Concentrate requirements in LLR systems Concentrate requirements in LLR systems (million tons)

	Feedlot beef	*Intensive sheep*	*Veal*
basic feed	14.9	0.63	1.1
by-products	5.0	0.12	0.5
Total "concentrate"	19.9	0.75	1.6

High concentrate rations common to all LLR systems are a logical reaction to low concentrate prices. Alternative rations in which less basic products are used and with comparable livestock productivity, are indeed available but implemented only in few situations mainly due to economic reasons. As shown by Algeo, replacement of grain by roughage would reduce profit per head drastically.

Rangelands

An often claimed positive effect of landless ruminant systems is a reduction of rangeland degradation through lower animal densities and higher off-take rates. The assumed side-effects have been important reasons for subsidising barley in many WANA countries and for stimulating feedlot systems in many African countries, most of which failed. However, no convincing evidence has been found for this assumed positive effect, probably because total meat production levels are not fixed. Thus, even though an initial reduction in animal numbers may appear, animal numbers on rangelands can also increase as a result of an increase in demand for sheep for the feedlot system.

A more fundamental and theoretical question is whether LLR systems require more or less land, compared to a situation where the same amount of meat is produced in land-based systems, using less feed grain. Considering that in most regions grass productivity is around 1.5 - 2 times grain productivity under comparable conditions, differences are likely to be small but probably advantageous in situations where less grain is fed.

The exact outcome of such comparison cannot be assessed in general as it is highly dependent on local conditions, e.g. type, yields and quality of alternative feeds, feed ration composition. Ward made a comparison between the present feedlot system in the USA and a situation where animals were purely range fed. Not surprisingly more feed energy (about 20%) is required per unit beef if animals are purely range fed. However, this comparison ignores the possibility of using by-products and inedible grain as possible supplements which would reduce the estimated difference.

A major exception of the general picture is veal production, as this is based on milk powder rations, being a highly inefficient way to produce meat. The relevance of the above considerations is limited. Firstly, because the quality of meat from animals fed grain-based rations and veal is often considered to be much higher than the quality of meat from animals mainly fed grass; many LLR systems exist mainly because consumers are prepared to pay the high premiums. Secondly, a reduction in production from these systems in a specific country is not likely to cause a linear reduction in arable land, considering the surpluses of grain which already exist in the EU and North America, while the reduced demand for grain is likely to result in lower yields per hectare and more rotational fallow land.

Animal Genetic Resources

There are no specific dangers to animal genetic resources from LLR systems, except for the demand for good quality dairy cattle for the urban dairies. These animals are purchased from the rural areas and are normally slaughtered at the end of the lactation, resulting in a negative selection in the dairy herds, especially the dairy buffalo herds, in the rural areas.

Contaminations of LLR-products

There is much concern about contamination of livestock products, as a result of the use of feed additives and drugs, particularly in intensive systems such as LLR systems, and the negative direct and indirect effect on human health. Different types of contamination may be relevant, e.g. antibiotics, antibiotic resistant bacteria, heavy metals, chemicals, hormones and mycotoxins, each with its own problems.

The extensive use of non-therapeutic use of antibiotics to increase productivity to improve the immune status of large herds, has come under criticism, because of increasing antibiotic resistance in bacteria which becomes a danger to human health. Most antibiotics are excreted in a relatively short time, and no residues will appear in animal products if recommended withdrawal times are complied with. However, there is evidence that bacteria pathogenic to both animals and man can acquire multiple antibiotic resistance in the gut of farm animals and can be transmitted to man via food or direct contact by farm workers. Moreover, non-pathogenic E.coli can act as a source of resistance to pathogenic E.coli and salmonella, though in practice transfers are rare.

The extent to which antibiotic resistant pathogens are the cause of extensive non-therapeutic use of antibiotics is hardly known, because of difficulties in tracing where a resistant strain originates from. Resistance can also be due to therapeutic use, both in animals and in man. Only few cases have proved that antibiotic resistant bacteria in humans did originate from (non)-therapeutic antibiotic use in animal husbandry.

Because of the potential problems with antibiotic resistant pathogens, only some antibiotics are allowed for non-therapeutic use in animal feeds, but thus far it has hardly resulted in a reduction in the prevalence of antibiotic resistant pathogens. This is probably because no measures have been taken simultaneously in respect to human use, thus still providing a major input to the pool of bacterial resistance.

The main public concern about chemicals is related to organo-chlorines such as DDT and lindane, as these compounds are highly persistent. The presence of organo-chlorines in animal fat tissues (Table 4), where they are mainly deposited, is not only the result of their use against ectoparasites in animal husbandry, but also by their presence in animal feed.

Table 4. Levels of organo-chlorines in animal fat tissues in Central Europe. Levels of organo-chlorines in animal fat tissues in Central Europe (mg/kg fat).

DDT	0.02 - 0.25
DDE	0.02 - 0.1
DDD	0.01 - 0.06
Lindane	0.01 - 1.0
HCB	0.01 - 3.3
Dieldrin	0.005 - 0.02
PCB	0.01 - 0.41

In the latter case animals are victims of pesticide use in grain production. As with heavy metals, concentration of organochlorines in livestock products may be higher than that in the feed ration, because several kilograms of feed are used to produce one kilogram of meat. However, this 'biomagnification' is partly neutralised by detoxification, metabolisation and excretion of chemicals by livestock, as by all living species.

In most developed countries, levels are well below security levels as organo-chlorines are replaced by other less problematic ectoparasitic drugs, and also as they cease to be used in grain production. In other countries they are still used both in livestock and crop production so high levels of organo-chlorines may occur in livestock products. In India, for example, DDT levels up to 7.2 ppm in animal fat have been measured, compared with levels up to 175 ppm in pulses.

Several types of hormones are used in LLR systems (legally, illegally or uncontrolled) to increase weight gain, feed conversion rates and/or to change meat composition (less fat) such as:

— anabolic steroids (e.g. testosterone, progresteron, zeranol),

— beta-agonists (e.g. clenbutarol, salbutamol),

— corticoids (e.g. cortisone), and

— somatotropines (e.g. BST).

Each hormone has different characteristics: e.g. beta-agonists are difficult to detect because they have a very short half-life time. They may cause allergic reactions and irregular heartbeats in humans. Stilbenes (e.g. DES) are detectable for several weeks after application and have proven to be carcinogenic.

Moreover, monitoring systems, even where available, are frequently criticised as being insufficient. Therefore, generalisation is impossible. Due to the short half-life of most hormones, little residue is expected to be found in meat products. In the USA, not one sample analysed exceeded tolerance rates. In the EU no hormones are allowed, though scientific basis does not exist for the ban of most hormones. Still, in some countries up to 10% of the meat samples analysed did contain hormones mainly beta-agonists and anabolics. Illegal use is caused by the fact that several of the hormones not allowed for regular application in fattening, are allowed for medical or veterinary treatment of humans or animals. In several countries neither regulations nor monitoring systems are present. Therefore, the extent of hormone utilisation is unclear, but it might be assumed that hormones are widely used to achieve higher profits.

Heavy metal contamination of livestock products, mainly in the kidneys and liver, are rare. If incidents do occur (mainly with lead and arsenium), they are usually related to high soil intake, which is uncommon in landless systems, and contamination of feed during transport. Regular additions to feed of Zn and Cd do not add to these problems.

Mycotoxic problems, such as aflatoxin, are not of major public concern, in contrast to what scientists believe. The biological effects of mycotoxins include liver damage, and nephrotoxic, neurotoxic, mutagenic, carcinogenic and teratogenic effects. Their impact on livestock is of prime importance; humans can also be affected but it is highly unlikely that livestock products contribute to this.

Mycotoxic problems are not considered as problematic in developed countries, mainly due to rigorous screening of feed samples by feed mills, though incidents with high animal losses do sometimes occur. In developing countries mycotoxic problems are likely to be much higher as storage conditions are much more problematic, and where monitoring of feed quality scarcely exists.

Note, data mentioned are not specifically related to LLR systems, as no such division is utilised in monitoring programs. Moreover, most aspects are not specific to landless systems or even livestock systems, e.g.:

- — improper use of antibiotic is still a major problem in situations where educational standards are low, veterinary services are scarce and antibiotic products freely available in the markets;
- — contaminations with heavy metals are more common in wild animals and fish than in domestic

animals, probably related to higher total feed quantities consumed by these animals as they generally become older.

The problems that are relevant to landless systems are mainly related to emergency slaughtering, to incidental contamination of feed during transport and storage, and to the complexity of the feed chains in landless systems. Particularly the last-mentioned problem is the most important: monitoring systems are often better developed in intensive systems, but screening for every possible contamination is highly labourious and costly. In some LLR systems, monitoring systems are not well developed, thus increasing the risk of contamination.

Fossil Energy Utilisation

Intensive livestock production systems are characterised by a high fossil energy utilisation. Nearly all research on fossil energy dates back to the seventies s and early eighties, but since then attention has declined with the reduction in fossil energy prices. Now, awareness is growing mainly because of the importance of CO_2 in global warming: the increase in CO_2 is expected to cause about 50% of the global warming occurring in the next half century.

Most comprehensive estimates on fossil energy utilisation for beef production amounts to 43.9 MJ per kg of edible beef, i.e. about 18.5 MJ per kg LW. Most of the fossil energy requirements are related to fattening in feedlots, accounting for 63% of the total requirements, mainly for feed production. As these estimates are mainly based on data from the seventies, energy efficiency may be improved somewhat by now. These improvements, however, are likely to be small as the improved FCR is already accounted for, while fossil energy utilisation in crop production has probably not decreased since a major part is related to fertiliser use and irrigation. Although no estimates are available for the other LLR systems on fossil energy utilisation per kg product, a few general remarks can be made.

Fossil energy requirements for veal production are likely to be much higher than for beef. Firstly, direct fossil energy use is higher due to the more sophisticated housing systems used, often including mechanical ventilation. Secondly, the fossil energy use for feed production is very high, e.g. drying wet substances to obtain milk powder. Brand and Melman calculated the fossil energy requirements for the Netherlands for 1 kg milk substitute as being 27.83 MJ. With a feed conversion of 1.5 the fossil energy requirements for 1 kg LWG would be 41.7 MJ!

Fossil energy requirements for sheep fattening are likely to be lower, mainly because of the better FCR and, to a lesser extent, also because less fuel and electricity is used in a direct form for farm operations. The fossil energy requirements for the production of one unit feed for sheep is not likely to be much different. The largest proportion is imported from developed countries and the part which is produced locally requires less fossil energy because of lower fertiliser use. However, this low fertiliser use could result in soil mining.

A major issue is how fossil energy requirements will change if livestock production becomes more land-based. The answer is highly dependent on local situations, e.g. alternative feed rations and transport distances. For beef fattening it is indicated that fossil energy requirements will reduce by roughly 16% if the fattening period in feedlots is reduced from 280 to 120 days per animal, while it is reduced by 24% if beef animals are all range fed. These reductions will, however, only have a small effect on global warming. The reduction of 16% for instance, will result in an effect on global warming equal to about 10 g methane per kg LW, compared to a typical methane emission from rumen digestion only of about 150 - 250 g per kg LW.

Waste from Processing of Animal Products

It is difficult to assess the contribution of processing to the LEI because:

(1) animals within LLR systems originally come from other systems and it is difficult to know which proportion of the LEI should be allocated to LLR systems and which to other systems;
(2) the emission factors from the different animal processing systems are highly variable and are largely dependent on the 'house keeping practices' of the processing plant; and
(3) the data regarding quantities produced and quantities processed is unreliable or unavailable..

Some tentative calculations are made based on statistical data from the LLR-system and on data of Verheyen et al.

Waste from Slaughter Process

Beef production in the USA: Total LWK (live weight kill), attributed to LLR systems, is estimated at 3.55 million tons to be processed in slaughterhouses. Average emission factors per ton LWK for the slaughter process are estimated at: 5 kg BOD, 0.3 N-Kj and 0.1 kg P in waste water. Emissions into the air cannot be estimated, as these depend too much on the energy sources used.

The total emission from slaughtering amounts to around 17.8 million kg BOD, 1.07 million kg N-Kj and

0.36 million kg P in waste water. Most of this waste water may cause major problems as slaughterhouses have a large capacity, but it is unknown to what extent waste water treatment is applied. Probably primary and secondary treatment is widely done, which mainly removes a fair amount of BOD, while tertiary treatment to remove nutrients is still to be implemented in most waste water treatment plants.

Veal production in the EU: Total LWK, attributed to LLR systems, is estimated at 1.08 million tons (disregarding the initial live weight of the calf, which is to be attributed to the land-based system). Based on similar estimates as above, the total emission from the slaughter process of veal in OECD countries, mainly the EU is: 5.4 million kg BOD, 0.32 million kg N-Kj and 0.11 million kg P in waste water.

Intensive mutton production in WANA: Total LWK, attributed to LLR-systems, is estimated at 0.22 million tons. Emission factors of the slaughter process in WANA should be estimated somewhat higher than for the USA and the EU, even though less meat processing is apparent. Per ton LWK estimates are assumed to amount to: 10 kg BOD, 1.0 kg N-Kj and 0.3 kg P in waste water. Total emission amounts to 2.1 million kg BOD, 0.22 million kg N-Kj and 0.06 million kg P. These figures are based on slaughtering in slaughterhouses. However, it can be assumed that a fair proportion of the sheep is slaughtered at home or in small butcheries. In that case waste production is difficult to estimate, and the figures are probably higher. Concentration of the waste water production may be lower in this case though the LLR-sheep-system mainly supplies cities, thus also this waste water production is likely to pose environmental and sanitary problems.

Waste from Tanneries

Beef production in the USA: The quantity of raw hides is around 8% of LW, i.e. 0.28 million tons. Verheyen et al. estimate the emission factors per ton raw hides from the tanning process at 550 kg solid waste, 100 kg BOD and 5 kg Chromium. Emissions to the air depend on the energy sources used during the tanning process. Further emissions to the air of volatile organic compounds are related to the finishing of the products and the paints used; these emissions depend on the end products.

Total emissions from the tanning of 284,000 tons of raw hides are around 156,000 tons solid waste, and waste water containing ca.28,400 tons BOD and 1,420 tons Cr. As tanneries are known for their waste water production it is likely that major part of this water is treated, particularly in OECD countries. In most cases, however, considerable part of the Cr is still discharged to surface waters.

No emission factors are known related to the tanning of sheepskins, but many problems around tanneries are common, sanitary as well as environmental..

Waste from Dairy Plants

Milk is produced within the LLR systems in the urban dairy system. Most of the milk is traded directly from here without further processing to consumers and shop keepers, and only a small portion is processed in factories. The waste in dairy plants in this system is negligible.

Biodiversity

LLR systems mainly affect biodiversity indirectly. Direct negative effects occur locally and are related to emissions of nutrients and organic matter to surface water. Indirect effects include those as a result of the production of animals and feed for the LLR systems. The effects on biodiversity of rearing young stock and breeding animals, mainly on pastures, will be described in the impact studies on grassland based livestock systems in temperate zones and (semi-)arid tropics and subtropics. Effects of producing feed are mainly related to the conversion of potential nature area into land for forage and grain production, which has an enormous negative impact on biodiversity.

One of the principal questions is to what extent are total land requirements increased by livestock production in LLR systems and compared with what. Land requirements of the present situation are probably only slightly higher compared to a system in which less grain is fed. However, the present requirements are substantionally higher when compared to a situation where ruminants mainly utilise "waste" biomass, as the positive effect of livestock production on human carrying capacity is only revealed at relatively low levels of livestock production.

7

Women's Role in Livestock Production

The involvement of women in livestock production is a long-standing tradition all over the world. Livestock patterns differ widely among ecological zones, and socio-political systems. Livestock production systems can be divided in four major categories. Nomads or transhumants, agropastoralist, intensive crops and livestock, and peri-urban intensive systems. In addition, there are some not-so-obvious livestock systems. In developing countries, the majority of livestock raisers are agropastoralists, deriving their incomes from both livestock and crop production. Agropastoral systems refer to a wide range of production systems, from the semi-nomadic to the sedentary.

The difference between agropastoral and intensive crop and livestock systems is that the former consists of larger herds and usually relies on some kind of communal pastures or rangelands. Agropastoral societies in Africa have in general more total numbers of livestock than in transhumant ones, but the livestock ownership per capita is higher among transhumants. However, there are many exceptions to this rule. Intensive crop and livestock systems are more popular as land shortages force agropastoralists to intensify their production. Such systems have fewer animals per household than other categories, and often rely on fodder production or crop residues and by-products. In Asia, where often land shortages are severe, there are fewer transhumant and agropastoral systems. The typical livestock production system is a smallholding integrated intensive crop-livestock farm.

It is difficult to generalise about the typical role of women within a livestock production system, as it differs even on a regional basis. In transhumant systems in Africa for example, herding and management responsibilities for large stock are rarely assigned to women. But that is not the case in transhumant systems in the Andes of Latin America.

In many societies women are responsible for small stock such as goats, sheep and poultry, as well as for young and sick animals kept at the homestead. They are mostly involved in milk production, although not all women control the sale of milk and its products. Involvement in this task is not necessarily the women's choice, but provides an opportunity to obtain some additional income within the given circumstances.

Production Systems and Function of Livestock

Livestock activities are normally integrated into the existing farming systems: animals graze on fallow land and browse on hedges, utilise crop residues as feedstuffs and produce milk and meat, manure for biogas and power for traction. A special category of smallholders are landless labourers, who own one or two dairy cows. This is a category that is dominant in Asia. These labourers and their families must also be considered and included in dairy development programmes. Experience in India shows, that through milk cooperatives a large number of marginal farmers, women and even the landless could be attracted to dairy production.

Sheep and goats can be kept on small farms without large fodder resources. They are a fundamental component in many farming systems, but they rarely dominate. In Niger most women own some sheep, which during the day are sent to pasture. Several studies show that this activity, when all costs for medicine, salt and straw are counted, provides little profit. Nevertheless, keeping sheep allows to women to realise some income, form part of their savings and is a source of prestige.

Poultry are probably the most important livestock species for many poor, rural families world-wide. Poultry keeping is largely the responsibility of women, but despite this, research into rural poultry development is usually narrowly focused on technical aspects with very little attention being paid to the wider socio-economic issues. Interventions to improve poultry production are often seen as a way to reach poor rural women to improve their livelihood. Project proposals that intend women to be the main beneficiaries should examine how changes will affect them and how much control they can exert.

A household is a unit of production where all members have the same objectives and interests. Experiences in Tanzania show that the husband and wife may have both shared and separate objectives or interests in dairy production, and each one tries to work toward achieving them. This situation has a great influence of the overall livestock management. Even if income is not the only factor that determines the socio-economic position of women, it greatly influences their status and well-being. Increasing women's income through improved livestock production would, therefore, also increase their status.

Men may feel threatened by this process and it is doubtful whether women would continue to maintain the traditional control they seem to have over the system. In fact, it has been reported when such changes take place, men will often take control. To avoid such a backlash, experience shows that projects must include men and women throughout all negotiations to bring about equitable and sustainable changes. However, at the same time efforts must be made to increase the capacity of women so that they are able to confidently negotiate and meet their strategic needs.

Risks and Best Practises for Livestock Projects

— The function of livestock for the various household members -men and women in particular- needs to be understood and fully accounted for

— Local practices and experiences are the basis for livestock production and further development has to take them into consideration, using technologies that are economically feasible, socially accepted and at low risk for farmers

— Special attention has to be oriented towards the role of women and their empowerment in the local and regional livestock production system.

— Traditional control over income within the household from cattle, milk, sheep, goats and poultry and the possible development under projects conditions has to be taken into account and possible results compared with project goals.

— In complex systems, special attention has to be given to the interrelations and the possible consequences of project interventions on gender.

For each specific situation, available information and relevant indicators will need to be considered together in the overall context of gender awareness.

Agricultural Production Systems

Livestock production systems and types of animals; crop/ livestock linkages; feeding; availability and quality of natural resources, ecological conditions and availability of land and pastures; soil quality; natural water sources; other common property resources; availability and cost of inputs; use of manure and crop residues; technology used.

Households and their Livelihoods

Role of livestock according to the men and women within the households; proportion of households with livestock and their social structure; ethnic, cultural and social relations; household activities and intrahousehold organisation; seasonal migration; relation between livestock and other activities; gender disaggregated seasonal occupation and sources of income.

Land Ownership

Insecurity of women's land tenure is one of the most serious obstacles to increase productivity of agriculture and livestock and the income of rural women. Land tenure refers to a set of rights which a person or organisation holds in order to own, have access to or use land. Security of land tenure is not limited to private ownership, but can exist in a variety of forms such as leases of public land or user rights to communal property. Tenure enables the holder to make management decisions on how land-based resources will be used for immediate needs and long-term sustainable investment.

Historically, in most cultures, women's access to land involved right of use, but not ownership. When common land is converted into state ownership and then to private land, women often lose their traditional rights and are often not considered when new laws are introduced. In addition, women are rarely aware of their rights. Generally, the importance of this overall scheme is neglected by policy makers, ministries and project designers. But there are also other examples. In Eritrea there is a strong gender equality concern in the land tenure legislation.

Since 1994 the right of ownership of all land is the exclusive right of the government. Every citizen, whose main source of income is derived from the use of land, has a lifetime right of usufruct over the land with the provision that such a right is neither divisible nor inheritable. Eritreans qualify automatically for land upon attainment of age 18 regardless of sex, religion or marital status. Individual holdings are registered and lifetime usufruc uary title deeds issued.

Due to their status within the family, in most societies men are the main owners of land. Private land is mainly transmitted from the father to the sons, and often daughters are only taken into consideration if no male successor is available. There are some exemptions, mainly in Latin America, where the land is divided between all children. A brief examination of the succession law provides an appreciation of possible future land distribution and subsequent farm size. If agriculture and pasture land is divided between all children, land availability per family tends to diminish rapidly.

With an increase of family members, in many cases, survival relying only on farming cannot be longer assured. In cases where there are no jobs available in other rural sectors to contribute to meeting family income needs, the family tends to increase productivity on the available land with mostly negative consequences on sustainability. Less fertile and exposed land is cultivated, crop rotation is no longer easible and the consequences are depleted soils, erosion and poor yields. If pressure on the land increases and no other income resources are created, migration is the last resort. In the first step men migrate seasonally, while the burden of the women and children who continue to cultivate the land increases. Abandoning the village and the migration of the whole family is often only carried out as an extreme second step, when there are no other options.

Security of land tenure is the key to having control over major decisions in agriculture and livestock production: what techniques to use, which products to sell and which to consume are examples thereof. The law of succession influences the distribution of land, the security of tenure and it is often a precondition for access to credit and a key link in the chain from household food production to national food security.

Some development projects have attempted to give women access to land. A World Bank project in India made it possible for women in Jammu and Kashmir to obtain joint title to mulberry gardens, if they have a letter of no objection from their husband or landowner. In Andhra Pradesh, state land grant schemes promoted women's access to land.

In a smallholder farmer project in Chile, obtaining land titles for female heads of households was a priority. This latter experience demonstrates that a government can successfully target the most needy farmers who lack secure land tenure and that rural women can be explicitly recognised as beneficiaries. Farmers with land tenure security are more readily to accept new technologies or interventions.

Risks and Best Practises for Livestock Projects

In many societies it can initially be difficult to establish who owns the land and who has the right to use it. But land tenure security is an important precondition for investment, as well as for empowerment. Given the complexity of different tenure systems, project strategy has to be adapted to the reality of the region and society.

In many societies women gain access to land only through a man. This reality cannot be rapidly changed and, therefore, project design has to adapt to this reality. Law of succession influences the distribution of land, security of tenure and is often a precondition for access to credit. Looking into the succession law allows project designers to obtain a better understanding of how gender relates to the possible development of farm size and pressure on land in future.

The concept of the household as a homogenous unit is inadequate. The power of a family member to make and enforce decisions depends on access to and control over land, livestock and other resources, as well as on access to income and social support networks. This needs to be taken into account in a gender perspective.

Ownership of Livestocks

Generally, men and women tend to own different animal species. In many societies, cattle and larger animals are usually owned by men, while smaller animals, such as goats and backyard poultry which are kept near the house, are more women's domain. However, ownership patterns of livestock are more complex and are strongly related to the livestock production system and to social and cultural factors.

Ownership of larger animals is often related to ownership of the land. How can a women own a cow while the land she uses belong to her husband? This question raised by projects in Africa illustrates the strong influence of cultural and traditional aspects. In the southern highlands of Tanzania, even if a married woman signs the ownership contract or pays for a cow, the animal still belongs to the husband, and even in case of divorce, the wife cannot take the animal with her.

Similar experiences are related from pastoral societies in Niger, where livestock is often a part of the dowry, but the control over the animals after marriage belongs to the man. The perception of these cultures imply that with marriage all the belongings of the women, including herself, reverts to the ownership of men.

The Nuer society in Sudan do not permit women to own cattle and goats, but they are often charged with the responsibility for grazing these animals. However, in extensive animal husbandry systems in Pakis an, women continue to own the animals they brought as a part of their dowry. They can decide by themselves what to do with them, but if they want to sell livestock, then they need the men's agreement.

Thus, even if women are the rightful and legal owners of livestock, they still depend largely on decisions and agreements made by men. In Burkina Faso, the handling of large animals is controlled by men, even if the women is the owner. In this society as in many others, for example Latin America, large livestock are held as an investment for savings and, therefore, are an important source of prestige.

The distribution of ownership of animal species between men and women depends not only on the society considered, but also on the type of animals raised. For example, in transhumant Peul society, each woman owns a cow to cover the family needs of milk and milk products. The more settled a family is, the more the division of ownership or control over different animal species becomes important.

Men tend to own mainly cows and camels, and women goats, sheep and poultry. But there are also exceptions to this rule. When the rearing of small animals such as pigs or poultry becomes a more important source of family income, then ownership, management and control of the animals is often turned over to the man.

Another way to look at ownership patterns is in terms of management of income generated from livestock. The general trend seems to be that men are the ones who control the income generated. But there are also exceptions to this. Examples from India show that women have learned to keep their own personal accounts and the pattern of income management in women-managed households is quite different from men. Generally, women's control over livestock resources tends to occur with widowhood and to increase with age.

Risks and best practices for livestock projects

— Livestock projects often assume that women ownership of cattle is a straightforward concept. Women buying or receiving a cow from the project does not necessarily mean in all societies that she owns it. More fundamental changes in society have to occur, and these changes are usually out of reach due to the time limitation of such projects. Analysing the specific situation in the project region, village and household, as well as monitoring changes, are important if one is to formulate and attain realistic project goals.

— Traditions are important, and therefore, the role of livestock as a source of capital, income or for traction and cultural ceremonies, etc., has to be considered. The perception of project planners of livestock-related priorities may differ from those of the target population and of the women in particular.

— As income earning opportunities in areas of livestock production traditionally controlled by women increase, their control may be taken over by men. Thus, an agreement with the beneficiaries, men and women, has to be found in order to avoid that the position of women is eroded.

Some indicators of ownership to be considered in the context of gender awareness are given below.

— number, ownership and control of livestock species by household types and gender;

— social and cultural constraint to ownership of livestock;

— role of livestock production (home consumption, commercial production, for savings, for ceremonies, for manure etc.);

— types of ownership and control of livestock products as meat, eggs, wool, dairy products etc.

Access to Capital and Knowledge

In a general rule, men have easier access to government provided credit than women. Women are rarely considered creditworthy because they have no collateral. In addition, they often cannot read and write, and are not used to frequent governmental or official institutions without their husbands consent and being accompanied.

In many countries, however, women have developed their small credit/loan systems. Credit funds and revolving savings of women's groups are common in

West Africa. The members of the group save a certain amount of money which is then granted to one of the women as a loan. Normally no interest is paid, and the social control guarantees that loans are repaid. Other credit systems consist of loans of animals or even milk for processing. Generally, these systems only function at the village level, often between neighbours, where social control can be assured.

Project experiences show that special credit lines for women are successful if they are transparent and cultural and social reality of the concerned families are considered. However, the local situation has to be well analysed. For exàmple, if the loan agreement or contract is only signed by the woman, but the loan is actually used by the man, or if jealousy or distrust develops between the two, then problems can arise and repayment of the loan will not be guaranteed.

In the most countries in Asia, Africa and Latin America, animal husbandry services are mainly oriented towards men. Veterinary services and extension programmes and advisory services have been mainly designed by men for men. Extension personnel are often not trained to teach technical subjects to women or to react their specific questions. Due to limited resources in time and material, attention is first given primarily to men's animals. Extension work with women often requires special didactic knowledge and communication skills because women often speak only the local language or dialect and illiteracy is high.

Risks and best practises for livestock projects

— Efforts to introduce new technology which does not take into account existing knowledge of men and women are unlikely to meet with success.

— Failure to direct information to the person responsible for a given activity may result in no increase in productivity or even in stock losses.

— Credit lines have to respond to client's needs and their social and cultural values. Social behaviour and traditional rules of men and women have to be well considered and credit lines adapted to their special needs.

Some indicators to evaluate household and gender specific access to capital and knowledge are given below.

— credit systems and required collateral;

— intra-household decision making and control over loans;

— proportion of total income saved on annual bases and use by household type and gender;

— preference concerning investment of liquidity (bank, purchase of livestock, jewellery, saving groups etc.), informal borrowing sources (purpose, interest, repayment conditions) etc.

Division of Labour

Patterns of gender division of labour are location-specific and change over time. Although the most typical pattern of gender division of labour is that women are responsible for animals kept at the homestead, there are many variations to this pattern from non-involvement in livestock to the management and herding of large stock.

If new livestock activities are introduced, it is mainly males who decide on whether or not to participate. The intra-household division of labour then depends on household labour availability, the number and type of livestock, economic development of the household and estimated income out of the new activity. But in fact, many decisions in a family are joint decisions, although they may not be formally recognised as such (not admitted by households and communities for socio-cultural reasons).

In Mali for example, handling of large livestock is traditionally controlled by men while women are responsible for smaller animals. This traditional division of labour is changing as the monetary value of livestock and their products increases. Experiences in Tanzania show that men primarily perform those jobs related to income generation and control most financial decisions.

From the Orissa region in India it is reported that women perform all the day to day activities related to caring, feeding, cleaning, health and production of livestock. Theses activities performed by women may appear to involve low skill levels, they are, however, most critical to the survival, health and production of the livestock. Activities performed by men are occasional in nature, involve less time, energy and labour and largely occur in the public domain, outside the confines of the household.

These are activities such as vaccinations, deworming, grazing, purchase of fodder and medicines, and taking animals to the dispensary. Clearly these activities involve greater mobility, access to new technology and information, greater interaction with the market and the outside world. Despite this division of work, livestock production and management continues in India to be a household activity with flexible arrangements of work between women and men. Women's access to information and training in mod rn livestock management and dairying continues to be limited and even indirect, lowering their involvement and efficiency. In Latin America, administration and control of cattle, including the milking, are done traditionally by men while the women participate in grazing activities and feeding.

Animal traction might reduce the work load for women, especially in Africa, where much of the hoe cultivation is done by women. On the other hand, the increased stall feeding and in general intensification of production might increase work load, as it involves transport of fodder or water over larger distances.

With increasing migration and off-farm work by men, the workload of women subsequently becomes greater as they become involved in activities once considered as being exclusively handled by men. Also in situations of war, women often take over the work traditionally carried out by men. It is, however, common for women to perform men' s tasks, whereas, the opposite rarely occurs.

It seems that a change of traditional division of labour occurs only by need, in cases where external factors influence a society, for example, the introduction of new technology, new agriculture or livestock activities, pressure for migration, war etc. Bringing shifts in division of labour by projects implylong term interventions.

With respect to children, gender-roles in most societies become internalised at a very young age - girls are socialised into performing roles traditionally performed by women and boys take on the roles considered appropriate for men. These internalised set of roles also influences attitudes and thinking and are carried later into life, which is why it is so difficult to change gender related issues.

Risks and best practises for livestock projects

— Seasonal differences in livestock activities such as feeding, watering and milking have to be taken into account, as well as seasonal changes in the labour input of different household members and their relationship to other farm and non-farm activities.

- As water is needed for livestock, as well as for milk processing, the year-round availability of an adequate water source near the farm is an important issue. Water resources that are located far from the farm, even for only part of the year, lead to very high labour inputs, frequently for women and children.
- Including women in project activities does not automatically benefit them. Periodic analysis of labour, adjustment of the time spent by women, or labour-reducing measures could diminish the risk of overwork.
- Special attention has to be directed towards children, especially girls, when increasing livestock and domestic tasks which can impact their ability to attend school.

Some indicators related to gender division of labour are given below.

- different types of livestock;
- important crops and other activities;
- cultural restrictions to livestock related activities for men and women;
- seasonal variation in labour intensity;
- hired labour;
- daily time use by gender for productive activities and domestic tasks;
- intra-household organisation and distribution of work.

Role of Livestock in Household Nutrition

Generally, household nutrition level through livestock keeping can be influenced in three ways:

— direct use of products like milk and milk products, eggs,

— using the income from milk, manure or animal sale to buy food

— using manure to improve household food production like vegetable and other food crop production.

Monetarisation of milk economy leads resource-poor families to sell more milk. But if the money is invested in nonfood items or used to drink beer, household nutrition level will not improve. The assumption that a cow will always lead to increased household nutrition does automatically not hold.

Nutrition levels of families have improved wherever projects have given focus on nutrition education or have brought multiple packages of intervention to improve the livelihood systems of the household. In transhumance societies in Niger, milk and milk products are mainly controlled by women. As women are traditionally more aware of nutrition aspects, they tend to assure family needs first of all, through direct consumption of livestock products or through selling and acquiring complementary products.

Contrary to crops, animal products, such as eggs and milk, are produced throughout most of the year. Selling them provides a small but continuous income, which is more likely to be reinvested in nutrition than the income of selling a cow or a cash-crop. Project experiences in Bangladesh show that through poultry production women's income could be raised.

Expenditure increased in portion to the increase in income. Most significant increase was expenditure on food, followed by clothes, savings, animals and schooling. It is interesting to note that as women's savings became more important, there is greater female influence on decision making, with the result that more girls are sent to school.

Generally, increased livestock production can have a positive influence on the nutritional level and the well-being of household members. Increased income from livestock production may change the intra-household distribution and control over products and earnings. When higher production and marketing activities become more important, women often lose their control over products and income. The level of nutrition within the family may decrease if the animals from which the products are derived are sold and the earnings spent on personal necessities, without taking into consideration the household well-being.

Risks and Best Practises for Livestock Projects

— Due to differences in men's and women's use of income, increases in men's earnings from livestock-related activities may not be necessarily translated into improved household nutrition, whereas women tend to first increase household well-being.

— New livestock activities can have a negative effect on household food security, if women-controlled activities change to the advantage of males or if new activities are taken over by men and women's' control of assets and benefits decreases.

— Religion, traditional beliefs and values may restrict the consumption of meat, milk and other dairy products by certain categories of people.

— In areas where livestock activities are mainly oriented towards subsistence, the establishment of milk collection and marketing points for meat, eggs, etc., may lead to a lower consumption of these products within the family. Project designers have to be aware, that if the objective of the project is to increase the income of small farmers, nutritional and social objectives for vulnerable groups should be realised through special programmes.

Information and indicators on household nutrition can be difficult to obtain. Some possible examples of indicators are given below.

— Eating habits by households and gender;
— seasonal variation in food combinations;
— food quantity and quality;
— proportion of animal products in nutrition and meal frequency by gender;
— religious and traditional constraints in nutrition of specific groups (children, pregnant women etc.);
— sources of different foods;
— part of income spent on nutrition;
— systems of storage, processing and losses;
— intra-household distribution and control over products and earnings;
— main diseases;
— access to potable water.

Influence of Livestock Business

If livestock keeping is the major source of income, men become more responsible and in-charge of the finance. Often men control the income and use it as they wish.

However, experience has shown that money from live animals sales, mainly cattle, is often used to pay school fees or make major house repairs etc., whereas money from milk sales is used for minor expenditures like buying soap, kerosene etc.

In many areas, sour milk, ghee and fresh cheese are the most common milk products processed at household level. Usually these products are for consumption and if marketed women control the income received. In cases where there is no market for fresh milk, processed milk products are sold in small quantities. When marketing of milk and milk products becomes a more important income source, commercialisation is often realised by men.

In Peul, Touareg and Roumboukawa societies, selling of milk is exclusively a women domain, independent of the quantity sold. In Burkina Faso, the selling of milk products is realised by men when there are long distances between the village and market place. Money earned through milk sales is used to purchase necessary products for daily life that are not available in the villages. Experience from India shows that women tend to have greater control on the income from sale of poultry, eggs, milk and small ruminants.

The introduction of Operation Flood in India, the organisation of milk collection programmes and the establishment of milk cooperatives illustrates how market organisation influences marketing possibilities and income generation. Therefore, intervention in livestock production should always be accompanied by a prospective evaluation of the existing or potential markets. Future market possibilities, involvement of women and income from livestock activities depend largely on the focus and support given by the project to market issues.

In general, women tend to spend the money they earn from livestock activities on the welfare of their families. Income from livestock activities is also invested into diversification of agriculture, to buy animals and even to buy land. In many societies, the little income derived from daily milk sales is sometimes used by men for drinking. This continues to be an intractable issue in many societies.

Risks and Best Practises for Livestock Projects

— The national and regional market structure, policy, prices, services and marketing possibilities, determines whether or not a specific livestock activity is economically viable. Gender specific division of work in processing and marketing as well as marketing activities of men and women have to be analysed and activities adapted to the specific society.

— Intervention in livestock production cannot be sustainable if market issues are not considered. Measures to improve productivity and production only succeed if through the marketing of the products beyond home consumption and additional income can be assured. Gender specific market possibilities and mobility as well as the control over the additional income has to be considered.

— Women often have a more limited scope for mobility than men, depending on the family structure and the region. Limited mobility impacts and strongly influences marketing possibilities for women.

Some indicators relating to processing and marketing of livestock products are given below.

— sources of income, especially from livestock activities;
— intra-household decision making and control over income and expenditure;
— structure of expenditure by type of household and gender;
— organisation and control over processing at household and industrial level;
— system of commercialisation of different livestock products;
— policy, prices and services;
— infrastructure;
— livestock products and quantities sold by household members through different marketing channels and system of payment;
— seasonal variation on prices for fresh and processed livestock products;
— processing and marketing possibilities for increasing livestock production;
— restriction to mobility of women.

Training in Livestock Activities

In many parts of the world, women and men are involved in livestock production, but, compared to women, men have easier access to technology and training, mainly due to their strong position as head of the household and greater access to off-farm mobility. In most countries, research and planning activities in the livestock sector, such as breeding, handling, feeding and health care, are largely dominated by men.

Official livestock services are often controlled by men and extension personal are primarily men who are not accustomed or trained to teach technical subjects to women. Extension programmes and educational materials are mainly designed by and oriented towards men. Although in most societies all household members are involved in some way or another in livestock production, the decision making processes within the family and the division of labour for activities such as feeding, milking, health care, processing and marketing differs between regions, societies and households.

At present, in many societies, women's access to information and training in modern livestock management and dairying continues to be limited and even indirect. Successful training should be oriented towards those household members which execute these tasks. For example, in societies where sick animals are mainly treated by women, they have a knowledge of the symptoms and cures for animal diseases.

But if they have no access to training, progress in best practices and appropriate herding to reduce diseases is difficult. Therefore, where extension services are dominated by men and where women have little access to training due to socio-culturally-defined gender roles, men need to be persuaded to see the relevance and the benefit of training women. Only through a carefully planned gender approach can livestock production goals and successful training of women and men be achieved.

Projects should identify and consider specific socio-cultural conditions of women, their needs and time constraints. Mobility of women is often limited and illiteracy high. Successful training can only be reached if these restrictions are considered and activities, approaches, methods and materials adapted accordingly to meet the specific conditions. Quality gender training should be practical and situational. Resource persons should represent both males and females, and should have an affiliation with agriculture and livestock production. It is also important to consider the age of the resource person. Very young facilitators and presenters may not be taken seriously by the group.

Risks and Best Practises for Livestock Projects

— To increase productivity in livestock production, training should be oriented towards those persons directly involved in these activities. Depending on the society, special training material for women has to elaborated. Training should be oriented towards the specific needs which in some societies can only be reached through separate courses for men and women.

— Constraints to women's participation due to time and social restrictions, as well as content of training, have to be analysed and concrete measures adopted to provide better assistance to women.

— Training should be balanced between the development of technical and methodological skills, and creating a social awareness for putting gender strategies into action.

Some possible indicators for developing activities in gender awareness training in livestock production are given below.

— differentiation of activities in livestock husbandry mainly executed by women and by men and possibilities to increase productivity through training;

— cultural and social constraints to participation of women in public life and extension;

— existence and structure of other production support services including other projects;

— constraints to mixed training groups;

— gender disaggregated adult literacy level;

— time and mobility constraints of women.

Role of Farmers' Organisations

There is little information on experiences of farmers' organisations, their impact at the local and regional level, and how they influence and impact on gender related issues. Farmers' organisations can play a vital role in the livestock development process. Input-supply organisations may grow and become centres for services such as artificial insemination, bulls, veterinary assistance, milk collection and processing, and marketing of animals and products.

The experiences in Andhra Pradesh in India show, that the membership of dairy cooperatives is largely dominated by men. Dairy cooperatives offered opportunities to men from backward communities to have access to benefits, emerge as leaders and gain visibility. Women only achieved symbolic representation and their are none or little opportunities for them to assume positions such as a manager, planner or director. In Orissa state in India, it seems that participation in the cooperatives benefits both men and women in terms of marketing. But there is clearly no significant impact on increasing women's decision making power or on enhancing their leadership qualities.

In some societies where the participation in cooperatives due to cultural and traditional reasons is difficult or impossible, women create their own cooperatives. By doing so, capacity building and decision making power, as well as self-confidence of women increase. Nevertheless, in these societies women cooperatives can only be successful if the husband first agrees to his wife's participation.

Risks and Best Practises for Livestock Projects

— Farmers' organisations should be based on local initiative, with some help and encouragement from external sources.

— Women's participation is important, but cultural and socio-economic realities have to be taken into consideration. Where in public men and women are separated, it may be necessary to create a women organisation.

— Capacity building efforts through farmers' organisations should, beside economic benefits, raise the awareness amongst women, build leadership qualities, and help them to gain a positive self-image. Literacy skills should be enhanced because poor literacy levels are often a major factor limiting a woman's access information, credit and market operations.

— Well organised farmers' groups also have the function of channelling the interests of their members and making these known in political circles. They may influence agriculture politics, extension services and project development.

Some indicators relating to farmers' organisations are given below.

— existence of farmers' groups, associations, cooperatives;

— their goals, orientation, structure, membership, activities, influence;

— level of participation of women and their influence in the decision making process;

— existence of women's groups;

— relations between different groups;

— main constraints.

Implications of Gender Aspects for Project Design

Today gender aspects are an overwhelming concern in all countries and in all fields of social and economic life. Gender is defined by IFAD as: "the socio-economic and evolving roles and functions of men and women as they relate to and complement each other within a specific socio-cultural and economic context". Despite such a definition, gender is often misunderstood as being the promotion of women only. However, gender issues focus not only on women, but on the relationship between men and women, their roles, access to and control over resources, and division of labour and needs. Gender relations determine household security, well-being of the family, planning, production and many other aspects of life.

Gender and gender related aspects are often given lip-service by politicians and briefly noted in policy papers. However, to actually incorporate the concept of gender across all operational levels is more difficult. To give the support required to ensure that it can be properly applied and that some measure of success can be made, demands even more effort.

Concerning livestock development, there is a high level of agreement in the literature that socio-economics and institutional frameworks play an important role in determining who does what, and who gets what. Social and cultural norms dictate the division of labour and control over assets. Policy and institutional structures often restrict existing sources of support to women, particularly credit to acquire large ruminants.

Values, norms and moral codes embedded in culture and tradition have very strong influence on gender issues as they determine attitudes and the organisational set-up of the whole community system. Like culture and traditions, political, institutional and legal structures also change slowly. Hence, these latter factors often impede the implementation of gender balanced programmes. Therefore, in projects operating at micro level, the most important set-up to consider is the socio-economic factor. Social and cultural factors determine the possible margin of action of women and their activities. In cases where women are excluded from community meetings, have no access to education and training, and where their capacity to become actively involved is not strengthened, they will always be left behind.

Economic factors are the basis for change because with a greater economic independence, self-confidence and possibilities of upward socio-economic movement increase. If one is to achieve a broad-based impact with a particular intervention, then gender aspects cannot be looked at separately: all factors including political, institutional and cultural aspects have to be considered.

The distinction between practical needs and strategic needs is not clear cut and the complexity of gender relationships goes beyond these two definitions. Nevertheless, these definitions are clearly relevant for gender analysis and monitoring. New gender frameworks and, more importantly, flexible ones have to be evolved to address the dynamic reality of gender at the grassroots level.

Failure to take into consideration gender relationships leads to unsuccessful project activities, and the marginalisation of the disadvantaged sector of society and a large part of the agricultural workforce. Thus, understanding gender relationships and adjusting methods and messages to them is crucial for full participation by all sectors of the community.

Gender Awareness of Project Designers

For an increasing number of organisations, participation of men and women is becoming an important goal. Many project papers mention the importance of gender aspects, which are often included in the overall project goals. However, upon closer examination of the operational plans, one finds that gender aspects are often missing. As demonstrated by an example from Burkina Faso, gender aspects are only included in words, but not translated into actions in the action plan for rural financing and sometimes are misleading for publicity purposes.

Formulation of gender balanced projects and programmes assumes a high degree of awareness of stakeholders on gender issues. This includes not only the theoretical understanding of the concept, but also the willingness of the parties and persons involved to think permanently about their own positions and concepts. Gender training helps to achieve a better understanding of the concept, but gender awareness is also a personal position in daily life. This is the reason why gender training should take all opportunities to move towards addressing gender issues at the level of the individual person, organisation and institutional fields, culture, etc. Few programmes have had the possibility to attempt this.

Nevertheless, training itself is not a sufficient condition for gender transformation. It has to be integrated in a capacity building strategy including follow-ups and action-learning projects on a continued basis. Innovative interventions in gender issues need to be initiated in order to bring about breakthroughs in traditional division of labour and culture.

Special attention has to be however given to the social and cultural reality in a specific society. Project focus and approach in gender issues have to fit into general attitudes of beneficiaries. Activities related to gender issues always require to analyse the specific situation in order to establish a gender balanced programme. Forced promotion of women rarely lead to sustainable impact or even worse, antagonism between groups could strengthen social imbalance.

In Mozambique a goat programme which promoted women ownership of livestock was in the beginning rejected by men and women due to the fear to disrespect cultural aspects. Though a gender balanced approach including men and women and giving special attention to women headed households the programme turned to be successful including women ownership and with participation of women and men in meetings on livestock issues.

Most programmes include gender without a proper understanding and perspective of the complete concept. Understanding gender requires a high awareness of the stakeholders and a permanent thinking about positions, behaviours and reactions of all involved partners, field workers and beneficiaries, and women and men in a project. A gender perspective should be a cross-cutting issue for all stakeholders, being part of their function and responsibilities, and should not be delegated or put aside.

Implications for Project Design

Gender awareness cannot be learned in a one week training course and it also cannot be delegated to one person or section within an agency. It is an ongoing process which implies the whole society in developed, as well as in developing countries. But a stakeholder can influence the process by his/her own commitment in applying the gender approach at each moment and for each action. Gender aspects should be adapted to the specific project region and society.

In 1998 Intercooperation worked out a guideline for equal opportunities for women and men within the organisation and at project work level and a document on implementation instruments to support gender approach with the goal to integrate a gender specific dimension of analysis and implementation into the daily organisational and operational practice.

Gender-aspects should be an integral part of project goals and get carried on board throughout the log frames. It should be systematically and practically included in the operational plan by translating it into concrete activities and relevant indicators. Apart from activity oriented indicators, which show that a certain activity has taken place, performance indicators should be well defined in line with the objectives and expected results or outputs. Proper monitoring to capture small, but sensible changes in gender relations within and among households should be worked out in all planning stages.

With the integration of gender aspects, the project's scope expands as it is forced to dwell into the social, cultural and economic parts of the target communities. It will no longer be a livestock project, but will also deal with household dynamics and community anthropology. This is one of the major challenges when integrating gender aspects in a project. Major attention has to be focused on the fact that gender changes are very slow and, therefore, project goals should not be too ambitious in gender issues.

Gender analysis requires taking into consideration other factors which could influence the potential impact of a project and presents opportunities or constraints to project goals and activities. The reason for specifying these determining factors is to identify what can facilitate or constrain the project. The following factors have to be considered:

— make sure that gender is not an issue of mistrust and prejudice, but of creativity, inspiration and positive spirit for men and women.

— social and cultural factors (norms and traditions which influence the behaviour of men, women and children, organisation of the daily life of the household members, specific religious rules for men and women)

— economic factors (poverty level, inflation, infrastructure, income distribution and distribution among family members, etc.)

— institutional structure (government, extension, education, health care, funding agencies etc., and their gender approach in theory and practice)

— environmental factors (quantity, quality and availability of land by households and intra-household distribution, water, energy, etc.)

— political factors (power relationship, system of decision making, legal system, etc., and their influence on the relationship of men and women)

— demographic factors (migration, life expectancy, infant mortality, etc.)

— legal parameters (right to ownership, law of succession, etc.)

For successful livestock interventions the following factors have to be considered:

- — Production systems and function of livestock
- — Ownership of and access to land, capital and knowledge
- — Ownership of different livestock species
- — Responsibilities and division of labour
- — Role of livestock products in household nutrition
- — Processing and marketing of livestock products and household economy

The task for project designers and planners is to assess these factors, and determine whether and how they will have an effect on or be affected by the project. An analysis of the flow of resources and benefits is essential to understand how a project will affect women and men. Monitoring is important and should be carried out periodically to evaluate whether the project is achieving the expected impact and, if not, corrective measures have to be taken.

In the past, livestock projects have been designed and implemented mostly by men. The main focus was oriented towards production issues such as breeding, feeding and animal health. Only when project development was hindered, were institutional set-ups, market aspects and gender issues considered. A review of more recent livestock project documents shows that a clear reorientation of projects in Africa, Asia and Latin America occurred.

Generally, the specific situation of the project region and its society, as well as the market aspects, are becoming more important. Nevertheless, a real awareness of gender and its application is today the exception, rather than the rule. In many countries, for example in India, livestock extension services are mainly oriented towards men, which makes it difficult to implement a gender balanced programme. Also, research is mostly dominated by men. Until today, little research efforts have been oriented towards livestock and intra-household decision making and organisation

The planning team has to be gender balanced, well trained on gender issues and show a high degree of gender awareness. Experiences on gender issues from other projects, within the same country or region, should be analysed and taken into consideration. This is particularly important if projects are oriented towards other activities.

Participatory Approach and Gender Training

Participatory methods are often used to initiate and guide the process of joint learning. These methods help people to visualise the analytical process of identifying causes and effects, and their linkages serve as common points of reference, and help to mobilise communities into action. As illiteracy is very high in most developing countries, especially amongst women, symbols help to visualise discussions and to record final plans and achieved agreements.

All those who participated in the questionnaire process agree, that despite the use of participatory methods and improving facilitation skills, not everybody participates in and benefits from the extension activities. Some agricultural trials fail and messages are not adopted because women, though being active farmers, did not attend the extension sessions. Husbands did not repay the loans because they spent the money on other things and had discussed neither the loan nor the purpose of the loan with their wives.

Women's groups frequently are unable to market products because their husbands do not allow them to travel. There are many other examples that one can cite which illustrate the difficulties in achieving satisfactory results through participatory methods, if gender aspects are not actually integrated in the extension system.

In building up a gender sensitive programme, training is one of the most important components. The overall aim of gender training is to increase the awareness, knowledge, skills, and behaviour in relation to gender of all people. The training has first to be oriented towards the staff in order for them to acquire knowledge and skills to understand, explain and practice the concept of gender. Awareness raising should be carried out at all levels, from senior staff to junior staff over to the clients themselves. Proper approach and teaching methodology should be developed for each level.

Specific approaches should be developed which suit staff and client levels separately. The background of the target community should be well understood and the training components should be in line with the existing situation. This means that a global gender training syllabus is not appropriate. Rather, a syllabus must be developed which specifically relates to the context. Resource persons should both represent the male and female sex, and should have an affiliation with agriculture and livestock production.

Another aspect consists in the preparation of training programmes. The aim of the training and the content should be well thought out and developed. A suitable training environment should be identified and proven techniques used to facilitate participants learning by understanding. These are two very important points because in gender training individuals basically talk about their own lives, beliefs and experiences.

Thus, they need an environment and facilitators which will promote open and frank discussions in order to identify certain biases and views they may hold concerning gender. The sensitivity of the gender concept itself calls for participatory, open minded and flexible approaches, so as to involve all participants in the whole process in order to identify their own weakness in terms of behaviour, perception and attitudes, and gradually prepare oneself for a change.

8

Technology in Livestock Development

In recent decades, the developing world has experienced a spectacular growth in the demand for animal products. A more affluent urban population uses a proportional or even increasing part of its growing income on animal products.While per capita consumption in the developed world remained stable, with some changes in between beef and poultry, because of health concerns of the first, people in developing countries have doubled their consumption of meat over the last two decades. Per capita milk consumption increased by more than 50 percent over the same period. Delgado and others introduced the term "Livestock Revolution" for this phenomenal growth in supply and demand.

Current per capita consumption in the developing world is only one-third to one-fourth of that in the developed world, and is likely to follow past developed-country consumption patterns, with a straight relationship between income and animal product consumption to about 70 kilograms of meat and 150 kilograms of milk per person per year. Considerable growth in demand for these products can therefore still be expected. This is also reflected in the various projections (Food and Agriculture Organisation [FAO] and International Food Policy Research Institute [IFPRI]), of which the work done by Delgado and others; Delgado and others; and Delgado and others at IFPRI are the best known. Delgado and his collaborators predict a further near doubling of the demand for meat and milk in the developing world, with the result that the developing countries in 2020 would out-demand the developed world in meat, and substantially approach the total milk consumption of the developed world.

This extraordinary growth in demand will be met almost exclusively by production in the developing countries. There are a number of reasons for this trend. First, the perishable character of animal products and the resulting high costs of transportation induce production to be located near the consumers, that is, around urban areas of the developing world.

Second, particularly in pig and poultry production, modern technology in almost all areas (breeding, feeding, and housing) can be easily transferred from the developed world to the developing world. Large units in the developing world now demonstrate similar or even better productivity performance indicators, such as growth and carcass composition, and efficiency indicators, such as fertility and feed conversion, than those in the OECD countries.

For example, feed conversion by poultry in Brazil, at 1.93 kikograms feed per kilogram growth, is lower than the 1.95 in the United States. In dairy production, with a much greater influence of physical factors such as climate and soils, and lesser economies of scale, modern technology is less transferable, and production has remained smallholder or family-farm based.

Third, intensive operations in the developed world face increasingly stricter environmental and animal welfare standards, and therefore find it difficult to compete with producers and processors from the developing world, with often less stringent and more loosely enforced standards.

Table 1. Past and projected growth (percent/year) in meat and milk production in different regions of the world

	Meat		Milk
Region	1983-94	1994-2020	1994-2020
China	8.0	2.9	3.2
Other East Asia	5.0	2.4	3.9
India	3.7	2.8	1.6
Other South Asia	4.8	2.6	3.1
Latin America	2.9	2.2	2.0
WANA	3.9	2.5	2.6
SSA	2.1	3.4	4.0
World	3.0	1.8	1.6

Growth patterns within the developing world differ significantly, with an almost explosive growth for meat production in East Asia in the 1980s and 1990s. Current and projected future growth of production is still significant, although leveling somewhat. Milk production gets an increased impetus in East Asia. Export-led growth is still particularly strong in poultry production, as in Brazil, which still had an annual growth in poultry production of 9 percent during 2000-04.

Structural Changes

These regional treads are leading to a number of global "hotspots" in East Asia and Latin America, where the strong growing production, in turn, has led to the following patterns of production in the developing world:

Economic Concentration

The Livestock Revolution caused a major increase in pig and poultry farm size in the "hotspot" areas. Several factors, although difficult to difficult to attribute individually, contributed to this trend.

First, the level of investment for housing, feeding, breeding, and animal health technology seems to favor large scale enterprises, although field research by Delgado and others does not give an unambiguous confirmation of this logical hypothesis.Their survey on the cost of production in different size pig, poultry, and dairy farms in East Asia and India showed that, if family labor was included, large enterprises had about the same profits per kg product as small farmers, but higher than medium size farms. Total profits were, of course, much higher in the large farms. Excluding the costs of family labor and environmental externalities increased the profits per unit product for the small farms.

Second, the concentration in the input and processing sector, combined with vertical integration and contract farming, also leads to increasing farm size. Those larger integrators prefer to deal with larger production units, while for the producers, at least initially, contract farming increases profitability as shown by the above mentioned survey by Delgado and others. Middle-income countries increasingly follow the developments in OECD countries. In the processing sector, for example in the United States, four companies have 60 percent of the poultry market share. The percentage of pigs produced under contract in the United States went from 10 percent in 1993 to 75 percent in 2003. In the Philippines, six integrators cooperating in an association now have 80 percent of the broiler meat market, dictating prices and quality and food safety standards. In Brazil,by 1994, four companies produced 40 percent of the broiler meat and the two largest had 60 percent of exports. In 2004, five companies managed 85 percent of exports.

Third, the increasing importance of supermarkets, with their demand for uniform and safe products, has reinforced this trend. However, while the rise of the supermarkets in developing countries is impressive-from a sales volume of US$1 billion in China in 1995 to US$55 billion in 2002, for example; the rise has been slower for fresh products, such as pasteurised milk and fresh meat. For meat, in East Asia the traditional fresh market will most probably remain the main outlet.

Supermarkets have played a major role in the rapid expansion of further processed and longer duration products such as ultra high temperature milk in China, where supermarkets now distribute 70 to 80 percent of the milk. In Latin America and in East Asia, they often rely, at least in part, on foreign direct investment, and have introduced higher quality and food safety standards and more reliable supply, and are therefore also having a general pull effect on the quality of other sectors. Finally, the policy and institutional framework favor large-scale production units. In many countries, the large-scale commercial sector, because of its political power, continues to have special privileges. This has significance for several areas:

— Large enterprises often benefit more from tariff or fiscal incentives. For example, in the Philippines the members of the main large pork producer association have preferential access to breeding animals (3 percent) and imported feed grain (35 percent compared with 60 percent for smallholders and independent producers).

— Large and influential enterprises also often have easier access to subsidised credit, which can promote inefficient, large-scale pig, milk, and poultry production in the peri-urban areas of developing countries.

— Per unit of product, large enterprises produce more waste, but, because of poor enforcement of environmental regulations, spend less on mitigating these effects.

Large-scale, intensive, meat-producing enterprises are therefore the main source of growth in meeting the increased demand for meat in the developing world. They buy a large majority of their stock feedand are thus completely divorced from the surrounding land base. Some figures demonstrate these trends:

During 1985-94, industrial meat production grew worldwide by about 4.3 percent per year, whereas small-scale production increased only by 2.2 percent per year. In fact, largescale, industrial production accounts for roughly 80 percent of the total production increase in livestock products in Asia since 1990. This is also demonstrated at the country level in East Asia and Latin America, although exact data on farm structure are often not available, and are not collected under uniform standards.

Spatial Concentration

Natural resources, such as climate and the availability of land and water, and labor, are traditionally important factors affecting business location. For perishable products, such as meat and milk, firms tend to locate near suppliers and clients to benefit from economies of scale and transport costs. For Western Europe, these forces have resulted in pig populations near port facilities

(Breton, Belgium, the Netherlands), feed production (Denmark), and urban conglomerates (Catalonia, Po delta). In the developing world, a similar picture emerges, as poor infrastructure (roads and refrigerated trucks) is causing intensive production units to locate near consumers, resulting in an excessive concentration of production along the Eastern Seaboard of China, in Thailand around Bangkok, and in Brazil in the State of Sao Paolo.

Some provinces on China's Eastern Seaboard, with the proximity of consumers and good port facilities, have a livestock density of over 500 livestock units per square kilometer, whereas on the basis of average crop requirements and arable land availability, a density of 100 would be in line with the absorptive capacity of the surrounding land.

Although livestock concentrations are increasing in the vicinity of urban areas, livestock expansion is also occurring in feed-producing areas, such as in the State of Mato Grosso in Brazil, where there has been a strong expansion of production and processing, based on the soybean and corn revolution. The extraordinary proximate concentration of people and livestock poses probably one of the most serious environmental and public health challenges for the coming decades.

Livestock Revolution

The extraordinary increase in demand for animal products in the developing world has led to a spatial concentration of larger-size production units mostly around urban areas, and economic concentration of processing and retailing, sometimes driven by foreign direct investments. This geographic and economic concentration in the livestock sector has significant negative-although also some positive-effects on the environment, public health, and social equity.

Environmental Effetcs

Livestock produce 12 billion tons of waste each year, and while most is recycled for use in crop production and energy, in several countries the amount produced exceeds the adsorptive capacity of the surrounding ecosystems and leads to the degradation of air, water, and soil quality, and losses of biodiversity.

Impact on ecosystems components

Livestock waste, through its nitrogen, phosphorus, greenhouse gas, and ammonia emissions, and heavy metalscan seriously affect surface and groundwater quality, soil chemical and physical characteristics, and air quality, and can contribute to global warming. These effects are detailed below.

— *Eutrophication of surface water*: Livestock waste, high in phosphates, nitrates, and organic matter, when directly discharged into open water, encourages the growth of oxygen-depleting plant life such as blue algae, and increases the turbidity of water, thereby preventing the penetration of sunlight and thus harming other aquatic organisms and destroying fish habitats. Nitrogen and phosphate from livestock waste are important sources of this process, which is caused by eutrophication. For example, livestock waste caused about 28 percent of total chemical oxygen demand of the discharge of the Pearl River in Southern China in 1996, and it is seriously threatening the mangrove, sea grass, and coral reefs stands of one of the world's most diverse shallow-water marine areas in the South

China Sea. This figure is expected to rise to 90 percent in 2010. Accelerated algae growth, known as "red tide", occurs with increasing frequency along the entire eastern seaboard of China.

— *Pollution of groundwater and soil.* "Nutrient loading" or use of feed, manure, and fertiliser nutrients in excess of livestock and crop nutrient demands, is a direct result of increasing animal densities and poor nutrient management practices. For example, 46 percent of the cropped area of East Asia is estimated to have an annual phosphate (P_2O_5) surplus of more than 10 kilograms, and already four percent has an estimated 40 kilogram per hectare per year phosphate surplus from livestock manure. Moreover, with chemical fertiliser added, a total of about 60 percent of cropland in China has a surplus of more than 40 kilograms phosphate per hectare per year. A similar picture emerges in the United States, where recoverable manure nitrogen exceeds crop system needs in 266 of 3141 counties, and recoverable manure phosphorus exceeds crop system needs in 485, or 15 percent, of 3141 counties. These high levels of nitrates and phosphates-saturate the soil, and therefore affect the quality of drinking water and soil structure and fertility, and hence the productive capacity of the land and the landscape.

— *Greenhouse gas emission.* Intensive production units-such as, cattle feed lots and intensive pig and poultry systems-produce fewer greenhouse gasses per unit of product than extensive production units. However, the intensive units still produce about 15 million tons of methane and one million tons of nitrous oxide (N_2O) annually.This is about five percent of total greenhouse gas emission. Methane emission is caused by enteric fermentation of ruminant (cattle, sheep, and goats) and anaerobic fermentation of large volumes of manure.

N_2O, produced by an aerobic process, is mainly the result of improper storage and inadequate separation of the different livestock waste components. Ammonia is the main cause of acid rain, which comes, in part, from the oxidation of nitrous oxide from manure, and in part by direct volatilisation from inadequate storage facilities and, in particular, from application to arable land. Ammonia affects aquatic life and forests, because it acidifies soils and water. For example, in the Netherlands, 70 percent of the country's nitrogen-related acid rain is estimated to come from livestock. Finally, ammonia is one of the main carriers of bad odors, which is one of the main friction points of intensive livestock production and society in general.

— *Habitat destruction:* The Feed Grain Connection. The Livestock Revolution will cause an increase in the demand for feed grainsfrom 657 million tons in 1997 to a projected 911 million tons in 2015, and 1,148 million tons in 2030. Delgado and others have similar projections. This means that at an average yield of about 4 tons per hectare, about 65 million hectares of agricultural land will be required during 1997-2030, or more than the total area of France, to meet the increased cereal demand. Earlier work showed, that most of these demands were met from supplies from environmentally less sensitive feed grain production areas, such as the American Midwest. However, with the increased scarcity of fishmeal for pigs and poultry production, because of the Blue Revolution, there has been a major expansion for in the demand for soybean meal, which in turn as caused a major expansion of soy bean cultivation in the humid forest areas of Brazil. Moreover, in the absence of major technological breakthroughs, which would either greatly increase feed cereal yields or the efficiency of conversion of feed into animal product, or both, the increased need for feed grains can be expected to lead to the conversion of sensitive biodiversity habitats, such as the tropical humid forest areas to arable land.

Public Health Effects

The public health effects of high livestock densities can be differentiated into direct human health effects of environmental pollution and indirect effects related to the closer interface within livestock populations and between livestock and humans.

Direct effects caused by environmental pollution

High levels of nitrates in water lead directly to Blue Baby Syndrome, with about 3,000 cases of infant mortality recorded over the last 40 years, and probably many more unreported cases in the tropics. More importantly, manure, through contaminated water or fresh produce, can carry a range of serious human pathogens, with high morbidity and mortality incidence in babies and children.

These pathogens, often asymptomatic in livestock, vary from bacterial pathogens such as E. coli O157:H7, campylobacter, salmonella, and leptospira spp., and protozoan agents (cryptosporidium) to viruses, such as the hepatitis A virus. While exact figures on the mortality and morbidity from water and food-borne diseases from livestock are scarce and attribution is difficult, water-borne diarrhea causes some 6,000 deaths daily, mostly among children under five years of age.Indirect effects Disease transmission from animals to humans (zoonoses) has occurred throughout history. Goats transmitting tuberculosis to humans, and cattle originating major measles epidemics, were earlier manifestations of such phenomena. However, more recently, there has been a particularly strong upsurge of livestock zoonosesand pathogens, which has brought this relationship to the center of public attention:

— Nipah virus in Malaysia in the mid-1990s caused 257 reported disease cases in humans, of which 115 were fatal, and led to the destruction of 1.2 million pigs.

— Bovine Spongiform Encephalopathy (BSE), indirectly linked with intensification, has been diagnosed in over 200,000 cattle in several European and North American countries, and the probable human expression in a new variant, Creutzfeld-Jakob disease (vCJD), has caused the death of 149 people in the U.K. and a small number of cases in other countries. The economic losses to this disease are estimated at US$20 billion.

— Severe Acute Respiratory Syndrome (SARS) was diagnosed in 8,101 cases in 2003 and 2004, with 774 deaths. It led to the death or destruction of about 120 million birds.

— Highly Pathogenic Avian Influenza (HPAI), as of 15 March 2005, had been diagnosed in 69 humans, with a mortality rate of about 70 percent. In Vietnam, for example, HPAI has resulted in an estimated drop in gross domestic product of 0.3 percent, and some estimates of losses already experienced by the entire region go as high as US$10 billion.

There are multiple causes of the emergence of new zoonoses, and include changes in climate and weather, land use and habitats, and human behavior and lifestyles; a growing population of more susceptible, elderly, and immuno-compromised individuals; and globalisation, which increases human-to-human contact and dietary diversity. However, the accelerating demand for animal products also plays an important role in the increased human health threat, because the increased geographical density of livestock and the closer interface between livestock and people greatly facilitate the jump of pathogens from one species to another, and the rapid spread within populations.

In most cases the pathogens originated in wild animals, such as the fruit bat transmitting the Nipah virus, civets in the case of SARS, and probably wild birds in the case of HPAI. In Nipah and HPAI, the pathogens were first transmitted to domestic animals, that is, pigs and poultry, respectively, before they became a threat to human health. Even in the case of BSE, intensified animal production and increased interface between animal populations proved a major driving force behind the emergence of the disease. Recycling animal waste of a different species (sheep) was the main vector in the transmission of the pathogen to cattle, and the use of new slaughterhouse technologies was the main cause for the transmission of the pathogen to humans.

At the animal level, virus introduction from wild animals is most likely to occur in extensive, open, smallholder systems, and therefore areas with a high density of smallholders are more likely to contract the disease. For subsequent spread of the disease, larger units must be infected, where the considerable movement of feed, live animals, and humans in such medium-size units, and the lack of genetic diversity, make them easy targets. The highest risk of an outbreak of these emerging zoonoses is therefore in areas where smallholder and commercial systems coexist, which is confirmed by some initial explanatory analysis by Gilbert and others on the recent HPAI outbreaks.

The significant increase in farm size in the developing world could constrain the emergence of these zoonoses, because the animal-human interface is less frequent and better controlled.This has important implications for smallholder systems, as will be described in the section below on equity effects. Similarly, smallholder systems and live animal markets play a key role in the transmission to humans, because they facilitate close contact between different species and between disease carriers and humans.

In general, two patterns of transmission emerge:

— In the first pattern, the actual transmission of the pathogen from animals to humans is rare, but once it has occurred, human-to-human transmission maintains the infection, as shown in influenza A and SARS.

— In the second pattern, animal-to-human transmission is the usual source of human infection and human-to-human disease transmission is rare. Examples of pathogens with this pattern of transmission include the Nipah virus and, until now, HPAI.

One of the greatest threats in this area is whether HPAI, until now in the second category, will mutate (eventually with other influenza virus strains) to change to the first pattern, upon which a pandemic could develop.

Transmission of Non-Zoonotic Diseases. The dominant role of supermarkets and consumers, and the strong increase in farm size described above, have resulted in an extraordinary drive toward product standardisation, which, particularly in the broiler and pig industry, is leading to a major narrowing of the genetic base of these stocks, with the consequent reduction in resistance to disease outbreaks.

The increase in the size of farms has generally increased their investments in protection measures from outside infection, and, therefore decreased the number, of disease outbreaks. However, the spatial concentration together with the increased susceptibility has increased the speed of transmission and the impact of animal disease, as shown by recent outbreaks of Classical Swine Fever and foot-and-mouth disease in Europe and Latin America. Fast transmission of the virus within areas of high stock density led to large economic losses, such as the 2001 foot-and-mouth disease outbreak in the U.K. that resulted in about US$5 billion in direct and US$8 billion in indirect losses. This danger is still more acute in the developing world, where large livestock enterprises are being established where there are very weak veterinary and human health services.

Use of Antibiotics and other Feed Additives. The need to impose stricter disease control under high livestock density situations, and the additional 5 to 10 percent gain in growth and feed conversion of antibiotics, leads to the use of high levels of antibiotics. Use of antibiotics in feed leads to resistant strains of, for example the food-borne pathogenic bacteria Salmonella and Campylobacter, and results in resistance of these bacteria to classic treatment in humans.

The extent of this problem is difficult to assess, because part of the antibiotic resistance is also caused by overuse by doctors, but animal-product-induced antibiotic resistance is certainly a part of the problem. The use of antibiotics in feed is allowed in most developing countries, although the extent, level, and quality of the products are difficult to assess. Illegal use is also widespread.

Although no official data are available, CDC and the Union of Concerned Scientists estimate a total of 25 million pounds to be in use, or 30 percent of total use for non-therapeutic purposes, such as increased growth. The use of antibiotics as a growth promoter was recently banned in the European Union (EU) and New Zealand, although the ban has seen an upswing in the use of antibiotics for therapeutic purposes, which leads to the suspicion of leakage to non-therapeutic use.

Equtiy Effects

The social and equity effects of the livestock revolution relate to the impact of changes in the structure of the production and processing, to the public health effects, and to the affordability of livestock products.

Crowding-out effect

The threat of smallholders being crowded out of livestock production by large-scale enterprises is often mentioned, in particular in pig and poultry production, where large units would significantly benefit from economies of scale, and where buyers prefer to buy larger volumes of a consistent quality, which smallholders cannot supply. However, the evidence emerging from recent studies is mixed, and other factors, such as the particular characteristics of the sector and the strength of the overall economy also seem important. For example, in Thailand, the number of small pig farms fell from 1.3 million in

1978 to 420,000 in 1998, which could be interpreted as a significant restructuring with possible important social implications. On the other hand, in the Philippines, which also witnessed a strong increase in the industrial sector, the smallholder system, with about 3 million farmers, maintains itself in parallel with the industrial sector.

Differences in alternative employment opportunities between the Thai and Philippine economies might explain these different trends. Such a restructuring did not take place in the Thai dairy sector, but did occur in the dairy sector in Brazil, where the number of smallholders delivering milk to the 10 largest dairy processors declined from about 175,000 to 110,000 during 1997-2000.

The decline was particularly strong in the suppliers of the three largest processors, which could indicate that the crowding out of the smallest producers is led by the processors. However, no systematic research is available on the characteristics of the out-migrating farmers, or on their future livelihood and employment. Moreover, the stronger backward and forward linkages of the industrial sector would provide alternative employment for smallholders pushed out by the larger units. Research in horticulture, for example, shows that the overall employment effect of farm size expansion on the total food chain is positive.

Vertical integration and other associative forms

Vertical integration, with one firm controlling more than one part of the supply chain, and in this case, meat or milk processors controlling, through long-term contracts, the supply of raw material from smallholder producers, has been advocated as an important tool for securing the participation of poor rural producers in the expanded market. The experience with vertical integration is mixed.

On the positive side, Delgado and others showed that contract farmers in the Philippines and Thailand received more than twice the level of profits per unit of product than independent producers, because they benefited from forward contracts in a situation of falling world market prices. However, experiences in the United States and, for example, China, point to some negative sides, because farmers felt they lost independence, their margins were reduced and, in China, contracts were complied with only if independent market prices were higher than the contract price. Anecdotal evidence from China also points to weak compliance from farmers when market prices are higher than the contract price.

Cooperatives also have a mixed record in the developing world, ranging from the very positive experiences with the cooperative dairy systems in India and several OECD countries, to governmentdominated, unviable systems in many African and East Asia countries.Public health effects and the poor Indirectly, the public health hazards associated with pollution and animal concentration affect the urban and rural poor, particularly. With about 10 to 20 percent of the population of the key "hotspot" countries depending on surface water for their daily needs, they will be most severely affected.

Affordability of animal products

The main social benefit of the Livestock Revolution concerns the lower price of meat products from the industrial sector, because it improves the access of the urban poor to affordable sources of essential nutrients. Generally, prices of meat products, particularly of poultry products, have declined in recent decades and the projections by Delgado and others foresee a further decline of about three percent per year during 1997-2020.

At the same time, the strong expansion of the demand for animal products has caused only a minor increase in feed grain prices; maize, for example, is expected to increase by one percent per year during 1997-2020, after a fall of about 50 percent over the last two decades. Thus, the main effect of the Livestock Revolution on agricultural prices seems to be to stem the fall in feed grain prices, such that maize and soybeans will increase in value over time compared to rice and wheat. This would, however, not harm the poor, because the prices of maize and soybeans would be much lower than the prices in the 1980s.

Environment-Public Health Farm Size Interactions

While most interactions are direct, two specific multiple linkages need more attention. They concern the links between farm size and pollution and farm size and emerging diseases. Both have a direct effect on social welfare.

Environment and farm size

Large pig and poultry operations produce more nutrient discharge per unit product than small farm units, thus confirming the public perception of those large units as the main polluters. Moreover, there is some evidence from East Asia that small producers pay more per kilogram product to internalise the environmental costs than the large units do.

Public health and farm size

The larger farms, resulting from the Livestock Revolution, have larger investments in bio-security, reducing the livestock-human interface, and therefore the chance of emerging new zoonoses. The enhance bio-security also reduces the animal-to-animal disease transmission, although the increase genetic uniformity makes these animal populations more vulnerable.

Moreover, new diseases emerged in particular, where smallholders and larger enterprises existed side by side. In areas with a disease outbreak, large enterprises, often aided by Veterinary Services, therefore seek to eliminate smallholder livestock holdings, also under the new policy of the World Animal Health Organisation (OIE) which allows the status of disease freedom (and therefore access to export markets) to particular areas within a country, and even particular farms in an area.

Small farmers are therefore often hit the hardest when affected by an outbreak of a disease, such as the Avian flu, although smallholder livestock is, throughout the developing world, an important source of income. Disease outbreaks therefore lead to the loss of animals through death and destruction of diseased or suspected animals, often a sharp drop in price and to major restrictions of a permanent nature on smallholder livestock husbandry practices. For example, in East and Southeast Asia following the HPAI crisis, the keeping of scavenging poultry and ducks, and marketing of live animals were forbidden, although only unevenly implemented.

Technical Solutions

New technologies, financial incentives, and regulatory and administrative changes will be required to deal with increasing animal waste loads and protect the environment, human health, and equity. It is unrealistic to expect to reduce demand, as argued by some, because the strong preference for meat and milk of the population of most of the developing world, and their current level of consumption, make it politically and even morally impossible to deny them these preferred foods. While there should be some emphasis on the place of animal products in a balanced and healthy diet, continued and even growing attention to prevention and mitigation of the negative effects of the Livestock Revolution must be the main focus of public policy.

However, experience in Europe and the United States shows that no generally applicable "silver bullet" solution exists. Zoning approaches aimed at changing livestock density, such as extensification or intensification, can only be part of the solution. Extensification in areas with an already high animal population density yields little environmental profit, and, in the absence of profitable manure processing, a reduction in livestock population will be often necessary in those areas. Concentrating intensive livestock farms in industrial areas seems to be a possibility from a spatial viewpoint, but has been faced with cultural and ethical objections. Disseminating the pig population to arable farming areas is minimally profitable, because of the limited extent to which the environment can be used.

Mitigating Environmental Pollution

Livestock nutrient waste can be a valuable input into crop production, and in many locations of the world, crops and livestock continue to be operationally and functionally linked enterprises. However, as crop and livestock enterprises became more specialised, as livestock concentrated near the feed or the consumers, and as inorganic fertilisers and transport costs declined, the two sectors separated. As livestock production specialised, manure became an undesirable byproduct, and the notion of its intrinsic fertiliser benefit has been replaced with a "waste" cost.

The management of animal waste includes a series of fundamental activities: waste production, collection, handling, storage, treatment, and land application. The specifics of these activities differ by livestock type, operational features such as animal housing and the presence or absence of storage, and the spatial integration of livestock with crop production to enable waste nutrients to be effectively recycled and not lost to the environment.

Numerous effective technologies to reduce pollution from livestock wastes are available, and, although many have been developed for developed countries, they are mostly transferable and have been adopted successfully in middle-income countries. Farmer adoption of waste management technologies is closely linked to need, capability, and cost. Economies of scale play an important role and depend on farm size, but there are very few win-win situations, so most waste management technologies need other measures to be generally implemented.The main technology areas are:

Feed management practices that reduce animal waste, consisting of more closely matching livestock diets to animal nutritional needs, by increasing the digestibility and bio-availability of nutrients in feed. This maximises feed conversion into meat and milk and minimises nutrient loads in waste, and is scale neutral regarding farm size. Animal waste handling, treatment, and storage technologies are available, although these are most economical in large operations. Their main financial attraction lies in reducing labor requirements, and the production of energy. Land spreading of wastes and recycling through crops, if carried out with adequate attention to soil and weather conditions and the overall nutrient balance, can be one of the more attractive options. The recent development and use of manure management decision tools have proven effective in integrating effective waste use into soil-crop-climate-livestock systems.

Feed Management Practices That Reduce Animal Waste

Strategies to improve waste management need to start at the front end, rather than the back end, of the animal, focusing on the relationship between feed, animal performance, and nutrient excretions in waste. Two technologies have been proven to be win-win propositions in that they are both profitable and reduce negative environmental impacts of waste:

— Improved diet composition and feeding practices to reduce overall volume and nutrient concentration in the waste stream; and Water conservation strategies to reduce waste mass, thereby making waste more transportable.

— Diet modification strategies for dairy, poultry, and swine are are farmsise-neutral strategies that could benefit a wide spectrum of small to large operations.

Feed

Depending on livestock type and associated dietary practices, significant reductions in manure volume and nitrogen (N) and phosphorous (P) loads can be obtained through improved diet composition, as shown in the big reductions in Flanders. For poultry, assuring proper and well-balanced protein, amino acid, and total N levels in the feed can reduce manure N output by as much as 50 percent. Reductions in dairy manure N and P due to dietmanagement are somewhat less, but are still a ..t 15 to 20 percent for N and 20 to 30 percent for P. Pig diet modifications can reduce waste N and P loads by adjustments in protein type and level, including use of low-stachyose soy,the use of enzymes, modification of feed particle size, and feeding newly developed corn hybrids, which are low in phytate.

An important aspect of dietary nutrient management concerns reduction in the excretion of N and P forms that are most susceptible to environmental loss. For example, excess protein in dairy diets is excreted in urine, which can be converted rapidly and lost as ammonia gas, or transformed rapidly to nitrate, which is susceptible to leaching loss. Feeding mineral P excessively increases water-soluble P in feces, the form most susceptible to loss in surface runoff from manure-amended fields. Feeding phytase or low-phytate corn to poultry and swine increases feed P availability, which can dramatically reduce total P in manure, but cause a slight increase in water-soluble P.

Water

Although not often considered a diet component, the use of water as part of feeding and cleaning systems can impact the amount of waste produced and transported for land application. In several countries, the price relationship between water and labor has reached such levels that large farms revert to manual labor. For example, some recently built large dairy farms around Beijing have abandoned their automated barn scrape and water-based flush systems and use local labor to clean and remove manure from barns.

Animal Waste Handling And Storage Technologies

Numerous manure handling and storage systems are available and used on livestock farms, from simple ways to stack manure for short periods of time to glass-lined cement structures for long-term manure storage. Whereas anaerobic lagoons are the most common form of waste treatment on farms that use scrape and flush systems, aerated lagoons use oxidation to break down organic matter and separate liquids for land application and solids for composting. Although fermentation processes greatly reduce the organic content and pathogen load of the effluent, the major problem is the removal of heavy metals, for which more sophisticated equipment is required.

Most dairy and pig manure is now flushed from housing, and manure is stored in outside lagoons. The widespread expansion of flush and lagoon systems in the United States and East Asia was premised on labor efficiency and the notion that storage would facilitate calculation of manure nutrients available and allow for land application during favorable weather conditions and close to crop nutrient demands. However, anaerobic manure lagoons emit more methane than other storage systems.

Anaerobic lagoons accounted for approximately seven percent of global methane emissions in 1991, a figure that has likely increased substantially. Also, emissions of ammonia, hydrogen sulfide, and carbon dioxide from manure storage, and carbon dioxide emissions from compost are other potential negative environmental outcomes of some waste storage systems.

Storage costs vary between US$1 and US$5.50 per 1,000 gallon storage space. The economics of waste storage and the storage period length depend on herd size, climate, and soil conditions. A principal rationale for encouraging the adoption of storage concerns favorable economics because of reduced labor requirements and better timing opportunities of the stored manure for land application, and hence better balancing the stored nutrients with crop nutrient needs.

Because of the economies of scale, most small-scale livestock operations will not be able to afford waste storage facilities, and therefore rely almost exclusively on family labor. The frequent removal and landspreading of manure fits their labor supply, although the semi-treated nature might have implications for human health. Small-scale operations need low-cost alternatives to current practices, such as improved ways to protect manure during the practice of short-term stacking.

Table 2. Conditions and characteristics associated with manure storage

Storage Period	Conditions and Characteristics
Short term (3 months or less)	Warm climate, no long periods of saturated soil. Pasture, grass, and hay land available for spreading. Equipment, time, and labor available as needed for frequent spreading.
Medium term (3 to 6 months)	To accommodate short periods of saturated soil. May not be adequate for some annual crop rotations. Some pasture, grass, or hay land likely needed for spreading.
Long term (6 months to 1 year)	Provides greatest flexibility for spreading. May best fit timing of cropping operations. Provides storage from one irrigation season to the next. Most flexibility for scheduling custom spreading operations.

Biodigestion for Energy

In addition to nutrients, manure also contains energy that can be converted into fuel. Biogas is a mixture of mainly methane and carbon dioxide produced when manure undergoes anaerobic fermentation, and contains approximately 60 percent of the energy value of natural gas. Optimum anaerobic fermentation occurs between 35°C and 37°C, and biodigestion does not work without outside heat sources in cold climates. In large-scale operations, manure is added at frequent intervals to the digester, biogas is removed continuously, and semi-solid effluent is removed periodically. The decision to use an aerobic digester can be based on several factors: the type and amount of manure available, alternative uses for the manure, projected use for the biogas, regulations regarding the purchase of the excess energy, odor reduction potential, and effluent use.

Disadvantages of biogas include its explosive nature (although no more so than natural gas or liquid propane) and its hydrogen sulfide content, which can be very corrosive to electrical wiring and toxic to humans if inhaled excessively. In countries where energy is relatively inexpensive, the costs of building and maintaining anaerobic digesters for the purpose of biogas production are high relative to the replacement value of the energy produced. Over the past 10 to 15 years, many digesters have been installed and used successfully on small dairy farms in Southeast China. In these and similar areas, biogas is used by farmers for lighting, cooking, and heating.

Additional advantages of biogas include odor reduction and a reduction in manure mass without nutrient loss. The effluent after fermentation contains all of the nitrogen, phosphorus, and potassium of the original manure and is a good source of fertiliser. The cost-effectiveness of methane recovery and energy conversion is still modest, although it maybe becoming increasingly attractive in areas were livestock concentration, and therefore the supply of manure, is high enough, and alternative sources of energy sources are high. Because of their energy generation and methane emission reduction potential, biodigestion might be of interest for carbon trade, for example, under the Prototype Carbon Fund, as currently explored under the World-Bank- funded National Environment Project in Brazil.

Waste Recycling Technologies

Effective recycling of manure nutrients through crops is the linchpin of proper manure management. The use of manure as a source of fertiliser N for crop production may become more attractive as energy costsincrease. Natural gas is used to produce a large percentage of inorganic fertiliser N, and natural gas accounts for 75 to 90 percent of the cost of making anhydrous ammonia. Furthermore, a reduction in the use of inorganic fertiliser N will reduce carbon dioxide (greenhouse gas) generation during its manufacturing. However, inorganic fertiliser is much easier to apply, and can be much more easily adapted to specific plant N requirements.

Decisions on the amount, form, timing, and methods of waste application focus on using waste nutrients optimally, reducing the movement of nutrients offsite, and preventing the delivery of polluted runoff to surface or groundwater to arrest the buildup of soil P and nitrate and the pollution of lakes, streams, and other surface water bodies. Emerging environmental policy in the United States is aimed at abating the emission of air pollutants from animal agriculture, which is also one of the main driving forces in the developing world.

Manure benefits crop production through its fertiliser value and enhancement of soil physical properties and overall soil quality. Two factors define a large part of its value:

— The conservation of urine N. Approximately 25 to 35 percent of the N contained in manure of ruminant livestock is available to the plant the season following application compared with 50 to 60 percent for poultry and swine.

— The difference in the N:P ratio of manure compared to the N:P requirements of crops. Applying sufficient manure to meet crop N requirements usually results in excessive P application, which can increase the hazard of soil P buildup and loss in runoff. The N:P ratio of manure can be aligned to the N:P requirements of crops by removing unnecessary mineral P supplements from animals' diets, through the conservation of manure N in storage, and during land application.

Controlling And Eradicating Intensive Livestock Production-Related Diseases

Livestock intensification has a direct public health effect through water and air pollution, and an indirect effect through the emergence of zoonoses. Technologies to address these effects are described below.

Table 3. Potential investments and inducements for waste management technologies

		Investments		Inducements	
Technology Domain	Key players in technology Implement-ation	Capital	Supplies& Services	Incentives	Disincen tives
		1	4	2	1
Feeding strategies	Producer, feed industry, research, extension /outreach	5	3	5	1
Waste collection, treatment&	Producer, , research, policy extension/				

storage	outreach				
Waste land application	Producer, policy, customized manure haulers, extension/ outreach	3	5	3	3

Note: Relative opportunity for investment and use of subsidies in scored from 1 to 5, 1 = low and 5 = high. a. an example of an incentive is cost sharing; b. and example of a disincentive is taxes or fines.

Reducing livestock-waste-related food-borne diseases

There is a wide range of treatment options available to reduce the microbial load. Storage systems using higher temperatures, and longer storage periods, reduce the disease threat of most pathogens, because they do not multiply outside their normal host. However, with survival times of more than one month formost pathogens, short-term storage alone is often not enough. Longer-term composting (at 55°C to 60°C), biodigestion, or aerobic treatment will significantly reduce most bacteria, but more drastic and expensive treatment systems using disinfectants, such as lime, or heat, will be required for complete safety.

Managing emerging zoonoses

Recent advances in human and animal immunology are enabling the development of diagnostic tools within a rather short time after a disease emerges, with the possible exception of Bovine Spongiform Encephalopathy (BSE), where only ex post diagnosis is possible. For example, several diagnostic tools for Highly Pathogenic Avian Influenza (HPAI) in humans, based on the detection of the virus by measuring the antibody response, were available in mid-2004, about six months after the outbreak of the disease in humans. Similar molecular diagnostics are available for animals, besides the normal microbial techniques. However, the molecular techniques require high-level laboratory facilities, whereas the normal microbial techniques are time consuming. There is thus a need for a rapid but simple (pen-side) diagnostic test that grass-root-level animal and human health staff can apply.

The control strategy for the animal vector of the most recently emerging zoonoses has been based on socalled "stamping out," where all diseased animals and those within a certain radius of the outbreak (normally 1 to 5 kilometers) are destroyed. Stamping out is often combined with a transport and live market ban. This has been the basic approach of the developed world, combining disease eradication with non-vaccination and strong import restrictions and quarantine.

Experience with this approach in developed countries has been mixed, as shown by the spread of the FMD and Classical Swine Fever outbreaks in Europe, which led to massive numbers of animals being culled, and massive losses, as described above. Experience with this approach in developing countries, where enforcement is weak and compensation inadequate, has been poor as well, particularly with smallholders.

As rethinking of the nonvaccination strategy is therefore recommended. Vaccine technology is less well developed and does not exist for, for example, the Nipah virus or BSE, or for its human variant, variant Creutzfeld-Jakob disease (vCJD). For HPAI, the fast-mutating characteristics of the virus complicate vaccine development, and only partial immunity is achieved with current available vaccines. A major issue (as with many other diseases) is the ability to differentiate between immunity induced by outside pathogens and by vaccination. If these effects cannot be differentiated, countries free of the disease do not allow export. Indonesia, with limited exports is therefore using vaccination, whereas Thailand has opted for continued stamping out. Differentiating infected from vaccinated animal (DIVA) vaccines are now being developed.

Similarly for FMD, although complicated also be the occurrence of several different strains, so-called "a marker vaccine", which would make it possible to differentiate vaccinated from pathogen induced antibodies, is practically available. Disease prevention, with strict biosecurity enforcement, often using special disease control systems (such as specific pathogen-free technology [SPF]) is now the most used alternative, particularly for large enterprises, but this approach is highly capital intensive, and anti-poor.

Environment and Public Health Technology

Most technologies described above depend on economies of scale. Public pro-poor technology development policies will therefore have to focus on research and disseminating those options that are scale neutral or have only a slight large-scale bias, such as feeding efficiency and vaccination, and leave the less-attractive technology approaches to the private sector, except those where there is a clear public good, such as greenhouse gas reduction.

Institutional Support Mechanisms

With few "win-win" technologies available and significant levels of externalities, public policy interventions are essential. However, public policy interventions are often focused on the easily observable effects, rather than on the most important environmental losses. For example, the links between pig production and air pollution are much more certain and direct than the impact of the land application of pig manure on water pollution. In particular, fewer policy options exist for controlling the impact of diffuse, non-point source pollution from agricultural production. Non-point discharges are difficult to monitor because they occur over wide areas and vary from day to day depending on weather conditions and the frequency and timing of application of potential pollutants, such as fertilisers and pesticides.

Moreover, there is still a lack of understanding of the relationships between "type of policy instrument-behavior of farmers-agronomical and environmental effects." The response of farmers to the implementation of manure policy and measures appeared to be more varied and complex than expected, as shown by the experiences in the Netherlands, where policy has gone through considerable trial and error .

Agricultural and environmental policy instruments can generally be divided into three main categories:

— Direct regulation or command-and-control instruments

— Economic or market-based instruments

— Communication or persuasive instruments.

Policy measures in the Organisation for Economic Co-operation and Development (OECD) countries have, in the past, focused on means (for example, a ban on manure application during certain time periods) rather than ends. The advantages are that such measures are relatively simple to develop, that the worst excesses can be dealt with, and a great deal can be seen to happen. It is probably also the most appropriate strategy in many developing countries.A move to a more targeted policy approach (for example, where farmers are obliged to achieve certain targets) has a number of benefits, because it gives farmers the freedom to select the most cost-effective approach to achieve the target practices and outcomes on their farms.

However, on the enforcement side, the major disadvantage is the difficulty in measuring the target. If a target cannot be measured practicably, (such as emissions to groundwater), a target derived from the original aim needs to be selected (such as mineral surplus). This is still only a proxy for the environmental damage that actually occurs.

There can also be spillover environmental effects arising from agro-environmental policy measures. For example, policies that place a limit on the amount of manure that can be spread can increase the quantity and distance over which manure is transported, and thereby the costs of manure disposal. Also, banning manure spreading during one season may cause excessive applications during another season and exacerbate pollution.

Regulatory Measures

Regulations are compulsory measures imposing requirements on producers to achieve specific levels of environmental quality, through environmental restrictions, bans, permit requirements, maximum rights, or minimum obligations. They leave no choice but to comply with specific rules or face penalties. They are the most common policy measure used in OECD countries to limit the environmental impact of livestock production.

Zoning is the most commonly used regulatory instrument. This can include the designation of nitratesensitive areas, restrictions on livestock numbers in certain areas, and the preparation and implementation of regional balances for nitrogen and phosphorus.

In the United States, concern has grown in recent years about the potential soil buildup and loss of manure nutrients to groundwater, lakes, and streams. These water quality issues have recently been joined by heightened awareness of the potential for livestock operations to emit pollutantsinto the atmosphere, which can adversely affect air quality and enhance nutrient enrichment and acidification of land and surface water resources. To respond to these concerns, federal and state agencies have increasingly focused regulations on the amount and timing of manure application to cropland. Farms designated as concentrated animal feeding operations (CAFOs) must now obtain permits for land spreading manure.

The current regulatory focus is on these large CAFOs under the assumption that they produce the most manure, and therefore pose the greatest environmental risk. However, it is becoming increasingly evident that farms of all sizes can generate negative environmental impacts, and that farm production components, such as barnyards and feedlots, may pose high environmental risk. In the EU, the regulatory framework has sought a reduction of overall manure production, rather than arriving at a better spatial distribution, because total livestock densities are so high that a better spatial distribution would provide relatively few environmental benefits.

There are three types of regulations placed on livestock producers that directly affect the level of manure production. Limiting livestock density. Norway was the first to introduce legislation in 1975 to limit the size of livestock operations. While not introduced for environmental reasons, they do reduce the environmental risks of intensive operations. Under these regulations, the maximum number of pigs for slaughter that can be kept is 1,400. In Germany, the number of animals that a livestock farmer is able to have is regulated by a maximum allowance of between two and three manure units per hectare. Limiting the expansion of livestock operations.

These regulations are in place at both the country level and in specific regions within countries. In Flanders, Belgium, the first Manure Action Plan banned new livestock farms. In Spain, the Restructuring Act stipulates minimum distances between farms and an upper limit on the size of new farms, making it difficult to set up a new farm of any size in areas that already have high pig populations, such as Catalonia.

Restricting the amount and way of land application of manure. These are in effect in many countries, primarily for the purpose of limiting water pollution. These restrictions vary from a set standard quantity across the whole country, to maximum levels established at each individual farm level, taking into account a range of input and output factors.

Most of the regulations relate to nitrogen, but some countries also impose restrictions on phosphates. To further limit water pollution, restrictions are often also placed on when manure can be spread and how close to waterways, ditches, wetlands, and so forth, to restrict nutrient runoff nutrients. These timing restrictions are generally stricter in countries with colder climates. Additional restrictions have been established in some countries to reduce ammonia emissions, involving the need to incorporate manure into the soil or restricting the manner in which manure is spread.

Regulations regarding the spreading of manure are also used to reduce the impact of odor air pollution.

Decision tools for regulatory policies

Advancements in geographic information systems (GIS) are rapidly increasing the understanding of the spatial and temporal relations among landscapes, weather, waste management, and environmental outcomes, and have become a critical decision tool for regulatory instruments.

Environmental management systems (EMS) developed in the United States focus on the farm level, enabling livestock producers of all sizes and locations to integrate information on environmental risks into existing farm management practices. The EMS framework helps producers evaluate existing facilities and management systems, and then identifies opportunities to take voluntary actions that reduce environmental risks, and ensure compliance with local standards.

The EMS has been tested widely on dairy, beef, and poultry farms in the United States. For the developing world the Livestock, Environment, and Development/ Food and Agriculture Organisation Initiative has developed a GIS-based approach, which overlays soil physical characteristics such as soil classification, groundwater levels, slope, flooding risk, and arable land areas, with other environmental parameters, such as biodiversity-sensitive areas, and commercial characteristics such as markets and transport infrastructure, with current livestock densities to define areas of surplus stock and areas with a potential for expansion. These technologies are now well developed, and have been applied, for example, in Thailand.

Institutional arrangements

Three important issues relating to the effectiveness of regulations emerge from experience.

First, there is the level and sector, in which the regulations are set and implemented. In the developing and developed world alike, there is a continuous struggle at the national level between Ministries of Agriculture, Environment, and Health (when public health issues are involved), on which institution sets the standards. The outcome of such a turf battle often determines the pro-production or pro-environment/health bias of the regulation. Moreover, there is the issue of central vs. local level, in both setting and implementation of the standards. Experience demonstrates the need for coordination between the ministries, with, in most cases, the environment institution, leading the standard setting, but coordinated at the cabinet-level. Standards can best be set at the national level, allowing for special standards for specific vulnerable of valuable areas, and the enforcement to be left at the local level.

Second there is the extent to which compliance with regulations is measured and assessed. Zoning and other regulations work best in the case of point source pollution (where the polluter can be unmistakably identified), where government has the financial resources to establish the infrastructure, and where there are reliable institutions to enforce environmental regulations. It will be more difficult to enforce, where institutions are weak, and where the polluter is difficult to identify (non-point source pollution), as in many countries in the developing world. More reliance has to be placed in those countries on market instruments.

Finally, there is the issue of rent seeking resulting from overly ambitious or unclear regulations, coming from the national level. Close involvement of local level, in particular in the implementation would therefore be needed.

Financial instruments

Because the livestock sector is an integral part of national economies, the macroeconomic policy framework in which the sector operates clearly affects its environmental impact. The European Union (EU) policy of preferential tariffs for cereal substitutes, such as cassava meal, clearly was one of the maindrivers of the strong increase in livestock density in Western Europe. The liberalisation of the economy in New Zealand in the mid-1980s, reduced the total nitrate emission between 1985 and 1995 by about 150,000 tons per year, although because the liberalisation caused a switch from sheep to cattle, it exacerbated nitrate surpluses in some regions.

Under increased global trade liberalisation, the same author predicted a reduction of nitrate overloads in the EU and Japan, and an increase in South America and Oceania, with an overall reduction at the global level. Specific economic instruments to mitigate environmental effects have not been widely used. In particular, environmental taxes and charges and tradable rights/ quotas have been implemented in only a few countries. Payments, particularly those relating to farm fixed assets, such as assistance in the construction of manure storage facilities, and on resource retirement have been increasingly used as a policy instrument. This has also been the approach in the few operations the World Bank has funded so far.

Payments based on farm fixed assets subsidise farmers to offset the investment cost of adjusting farm structure or equipment to adopt more environmentally friendly farming practices. Support has often been provided to livestock farmers to meet the requirements of regulations, particularly for costly manure storage facilities. The provision of this subsidy often varies over time, depending on when the regulations were introduced, and is often provided for only a limited period of time. For example, most of the support has already been provided in countries like Denmark, Norway, and Sweden, which have had more stringent storage requirements for a number of years. In addition to support for manure storage, some governments have provided financial assistance for on-farm capital investment in manure processing facilities.

Payments based on resource retirement subsidise farmers for retiring or removing resources from production, in particular from environmentally fragile land. Financing the exit of pig farmers was recently implemented in Belgium and the Netherlands. In 2000, the Netherlands introduced a package of measures involving the purchase by the government of manure production rights and pig quotas, farm audit arrangements, assistance to individual farmers, and demolition of farm buildings.

In 2001-02, the government bought out animal production rights and thereby decreased the number of pigs by about 10 percent, at the cost of about US$300 million. By 2002, the scheme had been instrumental in reducing the amount of phosphate in manure by seven percent and methane emissions by 6 percent. In Flanders, Belgium, a scheme was introduced in 2001 with the aim of reducing pig numbers by 50 percent. Farmers received a premium of almost EUR400 for every sow and EUR118 for every fattening pig they sold to the State. The government budgeted a total of EUR75 million for the scheme.

Environmental taxes and charges are sometimes levied on farm inputs or outputs that are a potential source of environmental damage, and have been levied on pig farmers in Belgium, Denmark,and the Netherlands, with the purpose of discouraging the excess production of nutrients in manure.

These are sectorwide taxes often covering all nutrient inputs, rather than those specifically from animal manure. While limited to just three countries, there has been an increase in the severity of these measures over time, in terms of both a reduction in the minimum threshold and/ or an increase in the tax rate.

Assigning quotas implies that the government administers and controls a set of production rights. Tradability of these rights promotes improvement of the structure and maintains the competitive power of the entire sector. Under tradable quota systems, the state assigns to farmers producer discharge quotas, permits, restrictions and bans, and maximum rights or minimum obligations, which are transferable or tradable.

In terms of livestock producers, such measures have been used only in the Netherlands, where the manure production rights of livestock producers, which were established in 1986, were made tradable in 1994. In order to reduce production levels, the government took 25 percent of the quota involved in each transaction. The system of tradable quotas (manure production rights) continued with theestablishment of a MINeral Accounting System (MINAS) in 1998. That same year, farm manure production quotas for pig producers were transferred into a tradable system of pig production quotas, based on the number of animals.

Communication Instruments

Communication instruments include collective projects to address environmental issues and measures to improve information flows to promote environmental objectives.

This information can be provided to both producers, in the form of technical assistance and extension, and to consumers, via labeling.

Most OECD and middle-income countries provide advisory services specifically targeted at improving the environmental performance of pig producers. This assistance can take a variety of forms, including technical advice regarding the construction of manure storage facilities, practical advice on the spreading of manure, the development of nutrient management plans, and the monitoring of environmental impacts. In the European Union, technical assistance has been provided to assist the implementation of the voluntary codes of good practice required by the Nitrates Directive. These inform farmers about practices to reduce the risk of nutrient pollution.

Community-based measures are those supporting public agencies or community-based associations (environmental cooperatives) in implementing collective projects to improve environmental quality in agriculture. Some governments have supported the development of alternative uses of manure to reduce environmental pressure and alleviate some of the constraints placed on farmers by restrictions imposed on the land application of manure.

There are a number of examples of voluntary environmental agreements between the agricultural sector and governments. Voluntary agreements between the agricultural sector and another private sector, such as the drinking water industry, are, however, rare. They can mainly be found in Germany and the Netherlands and to a lesser extent in France. Germany, with over 400, has the largest number of agreements. This is more than 80 percent of the total number of agreements in the EU.

However, in Germany, the regional distribution of cooperative agreements is unbalanced.

Apparently, a number of factors can promote or hamper the establishment of cooperative agreements between water supply companies and farmers in the EU. An important promoting factor is that drinking water stems from well-contained and compact groundwater resources, such as well-determined groundwater protection zones. This limits the spatial dispersion of pollution such that the cause of the pollution can be determined unambiguously.

Furthermore, the water supply companies' ability to finance the cooperative agreements, the willingness of the farmers to adopt pollution-reducing practices, and public preferences for pure and untreated water are supporting factors for the establishment of cooperative agreements. Important hampering factors are a reliance on command-and-control measures in some countries and a lack of enforcement of environmental legislation in others. Moreover, the existence of other local, regional, or national agro-environmental programs in some parts of the EU may crowd out local initiatives from water supply companies.

Impact of Wastemanagement Technologies

The complicated and long-term measurement requirements to ascertain cause-effect relationships between practices and water quality improvement may not be attainable during the relatively short term (five years). Water quality measurements are needed as baselines to assess long-term impacts. Environmental water sampling needs to be thought out in terms of what to measure where, when, and how often, to have the highest probability of associating measured water quality improvement to changes in farmer behavior.

Modest and, perhaps, more measurable shorter-term "proxy" indicators of environmental change should be considered. For example, if a project can show substantive gains in reducing manure nutrient loads through diet manipulation-and it has tested uncomplicated ways to track more equal distribution of livestock and manure nutrients and can shown the impact of storage on manure management-then the project will have much to say about adoption of management practices that will yield water quality improvements over the longer term.

Economics of livestock nutrient waste management

Little regional, or national, cost-benefit analyses of livestock waste management are available. Generally, they show that reducing nutrient loading and water pollution is more cost effective, than reducing those nutrients from surface water. For in example, figures from the Netherlands for 1998-2002 show that net loading of the soil decreased by about 0.2 kilograms P and 0.8 kilograms N per Euro spent. The cost of removal of N and P from surface waters is much higher.

More information is available on the costs and benefits of specific waste treatment technologies, such as bio-digesters, and manure as a substitute for inorganic fertiliser. Their cost effectiveness depends largely on the prevailing price of energy. In areas with low energy prices, such as in the US and in most developing countries, both bio-digestion and organic fertiliser have a doubtful financial profitability, and become only financially viable, when the environmental costs and benefits are included.Within these costs-benefit analyses, however, there is little information on the degree of public goods involved, and even less if these concern global, national, or local public goods.

Still, such information is crucial for assessing the possible incentives for private investments and the chance of scaling up, to be able to address these issues on much broader scale, than currently the case. Finally, internalising the environmental costs can also lead to major changes in the industry, for example, it has led to the complete phasing out of the pig industry in Singapore, and to major reallocation of intensive pig and beef operations from the Mid Western States to the South in the US.

Regulations To Mitigate Public Health Risks

Animal and public health regulations are set at the national level, by national authorities. In the area of control of zoonoses, the collaboration between animal and public health services is weak. Classically, the animal health services focus on animal-to-animal transmission, and the human health services on the human-to-human transmission, but there is a clear gap in animal-to-human transmission.

The World Health Organisation (WHO) and the World Animal Health Organisation (OIE) are responsible for facilitating the setting of health and food safety standards at the global level, in particular regarding cross-border trade, informing their members on scientific development and disease situations, and supporting national government services in disease control and eradication in the human and animal health sector, respectively. The recent outbreaks of animal-host-related diseases, such as the HPAI, have greatly strengthened the previous weak cooperation, as still exists at the national level.

Standard setting has been heavily dominated by the developed countries, which, in the animal sector, has led to the adoption of above described non-vaccination policies, seeking through strong import restrictions and quarantine to keep their own livestock populations free of disease. As experience has shown, in particular in Europe with FMD and Classical Swine Fever, but also in the Americas with BSE, this policy has not been successful, and a major rethink is required.

Weakening of the services has also affected their effectiveness. In real terms, budgets of veterinary services have generally declined, while staff numbers have increased, cutting substantially into the operating efficiency of the services. Moreover, the decentralisation policy implemented in many developing countries, including in the "hotspot" countries, such as Brazil, while certainly positive concerning decision making on local public goods, has weakened the capacity for fast, coordinated efforts to address issues related to national or international public goods, such as the control of emerging diseases. Delays in declarations of disease outbreaks and weak enforcement of quarantine and other measures are the result, as shown in the delayed reporting and slow start and unsatisfactory progress during the recent HPAI outbreak.

The ambiguous attitude of public services toward engaging private service providers in the early alert systems and early control is also one of the causes of the inadequate response capacity of public services. Privately operating service providers and grass-root para-health or veterinary agents are rarely used in early alert and subsequent control systems, and a coordinated approach between human and animal health providers is even more exceptional. As a result, the management of emergency disease outbreaks is poor in many developing countries, as shown by the survey of five countries on the experiences with the recent HPAI outbreak:

— *Disease surveillance and early alert systems.* The Avian flu was unofficially detected in Indonesia in August 2003, whereas the official declaration took place on January 25, 2004. This delay in officially notifying the World Animal Health Organisation (OIE) was likely common in most countries, as also shown by the proximity of the dates of declaration of the outbreak in the different countries, which is epidemiologically highly unlikely. This is partially due to inadequate diagnostic facilities and skilled staff, and partly of political pressure on the public health services to suppress this information, because of the economic consequences in lost domestic and export markets for their poultry products and tourism.

— *Stamping out.* Following the outbreaks, all governments adopted a stamping out policy, seeking to cull all birds in a 1-kilometer (Indonesia) to 3-kilometer radius (Lao PDR). However, implementation is deficient. In Vietnam the survey found that only 80 percent in the affected areas culled all their stock. Similar flaws were reported in the other affected countries. This was partly the result of inadequate (or totally absent) compensation. For example, Thailand, with the highest level of compensation, still paid only 70 percent of the value of the culled birds. Lack of awareness and poor enforcement are other reasons.

— *Movement restrictions.* Outside the stamping out area strict movement controls were established. However, even there, implementation was poor. In general it was found that 10 to 12 percent of the population did not follow the movement restrictions. For example, 8 of the 21 traders interviewed in Cambodia reported continuing trading, even experiencing a higher mortality, indicating the likelihood of diseased animals being traded.

— *Vaccination.* Because of international market access restrictions, only Indonesia decided to adopt a vaccination strategy. However, vaccination coverage was low, because of the difficulty of vaccinating free roaming animals and mistrust of the farmers of the quality of the vaccine. Vaccination coverage was therefore deficient.

These experiences show the inadequacy of the current system, and the need for innovation and action to be better equipped to handle either a likely resurgence of the current HPAI virus, or the emergence of new pathogens. They include:

— Making sure animal and human health agencies are independent to insulate them from political pressure to underreport disease outbreaks, which result from the current conflictof-interest situations between veterinary and commercial interests, in particular at Ministries of Agriculture.

— Strengthening public capacities in disease surveillance and early alert systems, and enhancing publicly fundedresearch on simple (pen-side) diagnostic tools and vaccines. Greater reliance on the use of vaccination, in combination with "stamping out". Ring vaccinations, (vaccinating the entire stock in a radius of up to 20-40 km around the disease outbreak) in combination with disease eradication, or regular vaccination strategies. There is a need to reassess the validity of current non-vaccination strategies of the OECD countries, in view of their trade, equity and ethical effects;

- Greater integration of private veterinary and human health operators in the surveillance and early alert systems.
- Regional quarantine (disease exclusion) schemes, thus rationalising the currently fragmented efforts along national borders, which are very difficult to control, and harmonising standards and procedures. The generally acknowledged principles in favor of a strong national disease management capacity also justify regional centralisation.
- Insurance schemes, at the regional or even global level, which would be able to pay acceptable compensation levels. Support for such a fund should be of interest to OECD countries, because it would directly protect their animal and human populations against diseases from developing countries spilling over into their territory.

The expansion of the demand for animal products offers potential opportunities for smallholders, but the economic concentration of the sector poses the danger that smallholders are crowded out. The overall equity effects of the Livestock Revolution on rural poverty are not clear, and need more study. Associative forms, including vertical integration and cooperatives, can provide some of the economies of scale, enabling smallholders to compete with large enterprises, but need to be built on solid legislation, which avoids monopolies and collusion of the integrators, and enforces compliance of both producers and integrators.

Rationale For International Involvement

Although the focus is often more on the "nuisance" factors (odor, flies) of intensive livestock production, concern about these negative effects is some middle-income countries. Several recently introduced stricter regulations regarding zoning and "end of the pipe" water quality standards, but the implementation and enforcement are weak. There is a genuine concern in several high-livestock countries regarding the need to strengthen surveillance systems, and to take early action. However, on the environment, whatever enforcement takes place is largely addressing these nuisance factors near urban areas and their effect on public health.

Serious attempts to arrive at a better spatial distribution of livestock, and to address the less visible and more long-term effects of high amounts of nitrate and phosphorous discharge, are still rare. Similarly, on the health side, the increased political attention to emerging diseases and animal and human health services has not yet been translated into increased budgets.

World Bank Involvement

The World Bank's support for better management of the environmental, public health, and social effects of the "Livestock Revolution" is still limited. On livestock waste management, the Bank has supported a number of initiatives.

The Livestock, Environment, and Development Initiative (LEAD), a multi-donor Trust Fund implemented by the Food and Agriculture Organisation (FAO), with active participation of the Bank. In livestock waste management, LEAD has developed and piloted in East Asia the tools for zonal planning and improved spatial distribution of livestock using geographic information systems (GIS) methodology and the calculation of national or regional nutrient balances. In addition, through its Virtual Center, it has focused global attention on this issue.

A World Bank loan and a Global Environment Facility (GEF) credit has funded the implementation of a livestock pollution mitigation project in Poland. The project seeks to help farmers in four counties meet the requirements spelled out in the European Union (EU) Nitrates Directive and in the Polish Law on Fertilisation. It provides a subsidy of up to US$10,000 per farm for the preparation of environmentally responsible farm management plans and the construction of tanks for storing liquid manure. Farmers contribute the remaining 50 percent. Performance in terms of tanks financed and nutrient management plans has been satisfactory.

Under GEF-World Bank funding the Danube Nutrient Reduction Plan in Serbia and Montenegro was launched, and under GEF funding, with FAO/LEAD, the East Asia Livestock Waste Management Project was launched, covering China (Guangdong Province), Thailand, and Vietnam. The emphasis in these projects is on mitigating nutrient loading, through the promotion of improved manure collection, storage, and application technology in selected watersheds, using matching grants as the most common incentive system. Policy and/or institutional needs are addressed in these projects through support for improving the regulatory framework and raising public awareness. The proposed East Asia Livestock WasteManagement Project also seeks to develop the tools and incentives to promote a better spatial distribution of intensive livestock production.

The National Environment Project in Brazil has given considerable attention to water quality as affected by intensive livestock production in the three southern states. Matching grants for livestock waste treatment (lagoons and so forth) were introduced, jointly, with a water quality monitoring system. A decision model to guide decision makers for licensing has been developed and the experiences are now being picked up by the private meatpacking industry. Finally, the project is supporting the construction of collective biodigestion installations for smallholders.

An additional interesting aspect of this project is its current negotiations with the World Bank's Prototype Carbon Fund to sell the N_2O reduction of biodigestion as carbon credits, in view of the greenhouse gas reduction of biodigestion. These interventions are introducing stricter regulations and subsidising technological improvements in manure management, and some introduce more innovative approaches. However, the more difficult areas of implementation and enforcement of these regulations receive less attention.

With a weak public enforcement capacity in many of the middle-income countries, and the prospects of large-scale subsidisation of manure management infrastructure rather remote, other mechanisms, focusing on market forces and "win-win" technologies and alternative enforcement methods on the basis of peer pressure need to be explored, assessing in particular experiences in these areas in OECD countries.

In the area of animal health, there has been some international support for the strengthening of public services, and the establishment of private-public partnerships, often as part of agricultural service support projects. However, the total amount from the developed world to the developing world amounts to around US$10-15 million per year, which is certainly not in relationship with the needs. In the specific area of funding the control of emerging zoonoses, the World Bank has supported the Government of Vietnam with an Avian Flu Emergency Recovery Project, a US$5 million credit for strengthening surveillance systems and rehabilitation of the sector, poultry sector infrastructure, and public awareness activities.

Public policy needs to play an important role because of the global and national public goods involved with the environmental, poverty reduction, and public health aspects. The World Bank, with its convening power at the global level, its capacity to combine national policy dialogues with the levels of investments required, and its cross-sectoral expertise covering, among others things, agriculture, environment, health, and infrastructure, can support these public policies. More details and the key entry points are provided below.

Global Public Goods

Several of the environmental and public health effects transcend national boundaries, as the environmental impacts affect global climate change, open waters, and biodiversity, and the emerging diseases, such as the Avian flu, have global significance. There is a direct interest of the developed world to assist developing countries in animal disease control, as their susceptible, high value livestock population, would suffer directly from any likely spill-over of these diseases from the developing world into their territory. While the adoption of the concept of animal diseases as a global public good, does not absolve the developing countries own responsibility, it would justify also more support from the North.

However, until recently, the global institutions concerned with research-such as the United Nations Environment Programmed (UNEP), the Food and Agriculture Organisation (FAO), the Global Environment Facility (GEF) and the Prototype Carbon Fund (PCF), the International Livestock ResearchInstitute (ILRI)-and the main environmental nongovernmental organisations paid little attention to the role of intensive livestock in affecting these global public goods.

Land-based production, integrating crops and livestock in smallholder systems, and pristine environments attracted more attention than the highly visible, ill-kept large facilities around urban areas. Groups from civil society that did focus on intensive livestock production were mostly linked to animal welfare agendas, such as was done by Compassion in World Farming (CIWF).

More recently, with the establishment of the Livestock and Environment Initiative in 1997 and the work of Delgado and others on the Livestock Revolution, there has been a greater appreciation of the challenges brought about by the global intensification of livestock production and processing. However, more needs to be done on global awareness, innovation, and sustainable institution development. This would include support for the following areas.

Innovation in Global Disease Control

As has been shown by the many disease outbreaks, in this era of globalisation, current approaches are not effective. New approaches must be developed with all stakeholders. Given the vested interests in this area, the involvement of an external, global institution-such as the World Bank-would be helpful to stimulate new thinking. As a first step, it is therefore proposed to establish a global platform for emerging zoonoses and other pathogens, supporting ongoing activities such as the World Animal Health Organisation (OIE) and FAO's Global Framework for the Progressive Control of FMD [foot-and-mouth disease] and other Transboundary Diseases (GT-TAD), but seeking to broaden it with private partnerships, from, for example, the pharmaceutical or processor and retailer sectors, and act as a think tank to promote innovative approaches for the control of these diseases. A proposal is being considered under the World Bank's Development Grant Facility to finance OIE for initial global awareness raising activities and additional studies in this area. These could include:

— Studies on the cost of "business as usual" in animal disease control, its implication for human health. Business as usual would imply maintaining veterinary services in the developing world at their current, under-funded and ineffective status, continuing current disease control strategies.

— Linked to the above study, a critical assessment of current disease control strategies, in particular the non-vaccination strategy, and an assessment of technology needs and costs and benefits of alternative strategies.

— Studies on the feasibility and eventual implementation of global or regional insurance and compensation systems.

— Support for development of models, in particular promoting regional integration of services, for the monitoring, surveillance, and early alert disease systems, for vaccination campaigns and for "stamping out" diseases, and imposition of quarantine measures, including adequate compensation of producers.

This platform would also serve as an advocacy instrument for increased funding, stressing the spillover dangers from the developing to the developed world under the current disease situation.

While there is a reasonable level of awareness under the livestock specialists community of the challenges of the livestock revolution, the major importance of these challenges are not yet recognised by the broader global community, and have not yet led to sustainable institutions and funding for the preventionand mitigation of the negative livestock and environment interactions. The following action are therefore required:

A greater effort in public awareness creation at global and "hotspot" country levels. Citizens need to realise that local actions that harm the environment can transcend national boundaries, creating negative impacts on regional and global scales. Sustainable, long-term integration of livestock-environment interactions is needed at national, regional, and global institutions.

Current programs of the main international organisations, such as FAO and the ILRI, have mostly a project-specific focus. The expansion of research on improved livestock waste management and health technologies. Global public support is required for those pro-poor or "orphan" technologies, described above, such as the increase in the efficiency of feed use, small-scale manure crop and energy recycle systems, robust and easy-to-apply disease and immunity level diagnostics and vaccines. The current level of funding for these areas is minimal.

National Public Policy

With national public goods of environmental sustainability of water, land, air, and biodiversity, poverty reduction, and public health at stake, there are major national and local public policy roles. There is a vast array of technologies, which for widespread adoption need to be supported by a similar expanded menu of financial, regulatory, and institutional instruments and support actions. The international community, including the World Bank should support:

— Greater awareness creation at the decision maker and general public level, on the major environmental and public health implications of increased livestock density, going beyond the "nuisance" factors, into the less obvious aspects, such as nutrient loading and emerging zoonoses;

— The preparation of national manure management plans in those countries with major current or expected problems, such as Brazil, China, Thailand, Mexico, Philippines, Vietnam, and several ECA countries. The preparation of such a plan would consist of major public awareness raising activities in the form of workshops and round tables, and a review of current and projected livestock population densities, their implications for the environment and public health, current experiences, for example within the framework of LEAD and GEF funded project, current legislation and policies, and their effect on livestock induced pollution. This would lead to the preparation of master plans for scaling up mitigation and prevention efforts of these effects. Scaling up should be gradual focusing on the sensitive areas, but start without delay;

— The development and implementation of regional zonal planning capacities, including the development of GIS technology, its supporting database, and the legislation and institutions to implement and enforce a better spatial distribution of livestock production;

— Strengthening the definition and implementation of market-based incentives for sustainable livestock waste management, including tradable quota systems;

— Strengthening national public animal and human health surveillance systems, in surveillance, early alert, and disease management capabilities, in particular through thepromotion of a closer integration with private grass-roots animal and human health systems, and a closer collaboration between animal and human health institutions;

— Developing associative forms of livestock waste management, such as cooperative biodigestion systems, watershed focused manure management plans, and so forth; and Developing and testing legislation supporting the development of supply chains, which adequately protect the interest of all stakeholders, including smallholders, and prevent the monopolies or collusion of the integrators.

The Livestock Revolution will continue, and could have major global negative environmental, public health, and social externalities. As seen, there is no "silver bullet" and experience from the developed world shows that there is not even a proven package of interventions, and trial and error approaches will be needed. Still, the threats are so significant that coordination among all stakeholders involved at the global and national levels is needed.

9

Sustainable Livestock-based Agriculture

Livestock production has been and must continue to be one of the most important economic and social activities of human culture. In these regions of the world, hundreds of millions of people depend directly or indirectly on livestock-based activities, the analysis of which is complex and multi-sectorial. Many technical and economic endeavours, at national and international levels, have attempted to increase animal production and animal productivity in the tropics but results in general have been meagre. The main reason is the lack of understanding of the ecological, socio-economic and cultural limitations inherent in these countries which constrain severely the application of conventional development models. There are also incredible opportunities for sustainable development, thanks to the enormous cultural and biological riches of the tropics, the rational exploitation of which could support sustainable production in the medium and long term, but which have not been considered seriously in previous attempts to develop the livestock sector in these regions.

The role of livestock in developing countries is quite complex and extends beyond their traditional uses to supply meat and milk as is invariably the case in the industrialised countries. They are certainly multi-purpose. They are valued for one or several of the following traits: capital, credit, traction, milk, meat, hides, fuel and fertiliser. Thus, for families without land, livestock are primarily a means of increasing the family income. For the crop farmer especially in Asia, but increasingly in Africa and Latin America, the large livestock—cattle and buffaloes—are primarily sources of traction and power. In many societies the dung is used for fuel and to a lesser extent as fertiliser. For the transhumant grazier, livestock may be most valuable as a capital resource and a source of credit. Production systems must take into account these varied roles, and must be adapted to specific local situations. If, as expected, fossil fuel prices increase in the long term at rates exceeding average inflation in the industrialised countries, then one increasing role will be the use of livestock as sources of power in agriculture. This is already the case for many countries in Asia with low GNP and low international purchasing power.

The other issue, which perhaps relates more specifically to Latin America and parts of Africa, is that the principal livestock production system is extensive grazing by large ruminants, the establishment of which has mostly been through the destruction of the natural ecosystems of the tropical rain and cloud forests. These systems have consolidated the position of the medium to large landowner/cattle rancher and, by so doing, minimised opportunities for the small-scale farmer. Despite the privileged role accorded to extensive cattle ranching—witness the supporting research and development efforts over more than three decades by both international and national institutions—these production systems have become increasingly less profitable, due to increased prices of animals, feeds and other inputs, as well as increasing land prices due to competition with other end use patterns. The result has been their conversion into secondary activities kept in place by support from industry and commerce.

Livestock are enormously important to the economies of the less-developed countries as a whole. According to Brumby, when the value of livestock in providing rural transportation, draught power for cultivation, manure for crop production and their ability to utilise non-arable land and the agricultural residues are added to the direct economic value of animal products, livestock accounts for about half the total agricultural production. Livestock also play a critical role in maintaining a cash flow for poor farmers who grow their crops essentially to provide food for their own household. Milk, meat and hides will always be sought after by those segments of society that have the necessary purchasing power to acquire these products.

Renewable Natural Resources

It is becoming a matter of increasing concern that the present rate of economic growth is already outstripping the capacity of the earth's ecosystems:

(i) to produce the required resources; and

(ii) to absorb the pollution caused by present levels of economic activities.

The impact of the expected doubling of the human population by the mid-term of the next century, most of which will take place in developing countries, coupled with the aspirations of the present and future under-privileged majority, poses a threat that can in no way be described or predicted.

It is quite clear that future scenarios of resource utilisation must be predicated on:

— Optimising the capacity of the earth's ecosystems to produce biomass, as the only renewable source of energy, chemicals and food, without compromising the biological diversity on which the survival of all ecosystems depends.

— Minimising waste through recycling, which reduces the need for raw materials and helps to protect the environment.

For livestock to play a symbiotic role in such a scenario, it will be necessary to give priority to species that combine efficiency of conversion and productivity, produce low emissions of methane, and have the capacity to use by-products and residues from other primary industries.

Pigs and scavenging poultry undoubtedly are the preferred animal species in this scenario, but there will be an increasing role for the small, as opposed to large, ruminants and for the small non-ruminant herbivores. High reproductive rate is what gives the competitive edge to these species.

Aquatic systems with multiple production of fish, ducks, geese, water plants and other animal and plant species will also find an increasingly important niche in the new livestock development model. The large herbivores will have a primary role as sources of power and fertiliser for agriculture, which they will achieve by recycling the residues from the crops they help to produce. Increasingly they will be expected to combine these activities with milk production and reproduction. The use of castrated males for work is a luxury which future pressures on resources will make increasingly less attractive. As with the plant kingdom, the need for biodiversity *per se* will justify the domestic use of the widest possible range of livestock species.

Among the largely unexplored possibilities of the diverse animal species present in tropical ecosystems, the natural wildlife—mammals, birds, reptiles, fish and crustaceans—can also make a contribution to sustainable livestock production systems, especially because of their adaptation to the ecologically fragile zones and their contribution to biodiversity.

Sustainable Use

The World Commission on Environment and Development defined sustainability as "ensuring that development meets the needs of the present without compromising the ability of future generations to meet their own needs". To this can be added the need to respond to the pressures increasingly coming to bear in both industrialised and developing countries to safeguard natural resources. In livestock-based agriculture, production systems must take into account these issues. In practical terms this means measuring the "sustainability" of the system according to its effects on:

- — the economy,
- — the environment,
- — the need for energy,
- — animal welfare, and
- — food quality and security.

Economic Issues

The prerequisite of any livestock system is that it should be profitable to the producer. In all industrialised countries, the costs of livestock production have escalated mainly because of the increase in the cost of labour caused by rising expectations and competition from other industries. The situation is exacerbated in those countries where farm size is small and therefore unit costs of mechanisation are high. Faced with such situations, governments have resorted to subsidising agriculture through guaranteed support prices and other forms of financial assistance. The total cost of this support amounts to a staggering 75% of the total value of agricultural production in Japan, 40–50% in the European Economic Community and up to 25% in the USA.

Producers are supported in industrialised countries through subsidies and protected markets. These supports have two important consequences:

(i) they increase the price of food to the domestic consumer; and

(ii) they reduce the economic growth of many developing countries unable to export primary and secondary commodities against the barriers of tariffs and quotas.

Such policies in the long term are not sustainable. They are inefficient in resource utilisation since they direct expensive resources into products which could be produced elsewhere with fewer resources. The production of wheat and milk in some oil-rich countries, which is only made feasible with massive inputs derived from fossil fuel is an example of this misguided policy.

For the world economy to grow at an optimum and more equitable rate, it is essential that there is free trade in basic commodities. The objective of GATT (General Agreement on Trade and Tariffs) is precisely to promote the concept of "comparative advantage", whereby commodities are produced in the areas/countries which use least resources for that purpose. Unfortunately, the free movement of capital means that the principal beneficiaries from the exploitation of "comparative advantage" are likely to be the large multi-national companies. An even more worrying issue is that comparative advantage can also mean advantage gained by not paying the environmental cost of a given production activity.

Another world trend likely to have considerable economic impact is the future cutback in the availability of cereal grains for livestock feed. Two factors will contribute to this trend. On the one hand, the rising human population in low income countries will increase the demand for cereal grain which usually is the cheapest staple either to produce locally or to import from world markets.

On the other, state subsidies and protection, although still at high levels in the industrialised countries, will gradually be reduced as a result of the GATT agreement. Grain prices will rise as a consequence. Increasing cost of agro-chemicals and fossil fuel, and environmental pressures, will lead to cutbacks in the use of these inputs which in turn will lead to lower crop yields and increased costs of production.

For the poor small scale farmer in a less-developed country where subsidies on the scale presently employed by the industrialised countries are out of the question, the priorities are food security and to maintain their life style. The essential steps to achieve this are to produce first for family consumption, using an integrated production system involving crops, forestry and livestock, and which ensures self-reliance by making maximum use of renewable natural resources with minimal dependence on inputs from outside.

This is a more economical and ecological way of improving their standard of living, as compared with the developed country models which have used fossil fuel to achieve this end.

Environmental Problems

The productivity and efficiency of livestock production per animal unit in the least-developed countries is considerably less than in the more-developed world. But 'productivity' and 'efficiency' are references that relate specifically to temperate agricultural practices. In the tropics, livestock activities are different—how does one measure the efficiency of survival, or as a credit institution? These and other productive traits are achieved with minimum inputs of fossil fuel.

The biomass availability and the potential to produce more biomass in those countries which are in the tropics is many times higher than in the major industrialised countries which are exclusively situated in temperate zones. But we have only just begun to recognise the potential of tropical feed resources, let alone devise ways of exploiting them in a way which will be sustainable.

Another factor likely to become increasingly important in the future is the potential for tropical soil-based ecosystems, derived from decaying biomass, to foster atmospheric nitrogen fixation, sequester carbon and oxidise methane. Threats to the environment come from:

- — atmospheric contamination,
- — deforestation,
- — accelerated erosion,
- — soil and water pollution,
- — loss of biodiversity, and
- — excessive human aspirations and lack of awareness of the finite nature of renewable resources.

Global warming

Livestock production is intimately linked with build-up of atmospheric carbon dioxide and methane since:

(i) emissions of carbon dioxide are caused mostly by burning fossil fuel and tropical deforestation;

(ii) some 20% of methane emissions arise from digestive fermentation in the gut of herbivores, the methane itself contributing to some 15% of total greenhouse gases.

Forests and high biomass producing crops are important sinks for carbon dioxide. Decaying biomass in contact with soil appears to be an important ecosystem where anaerobic micro-organisms oxidise methane. Use of animal traction reduces the burning of fossil fuel; and permanent tropical deforestation is mostly caused by activities leading to establishment of pastures for extensive ruminant livestock production.

Alternative methods of livestock production using high biomass-producing crops, fed mainly to monogastric animals and small herbivores, in partial or total confinement, will lead to increases in the size and number of sinks for both carbon dioxide and methane.

Deforestation

Extensive cattle grazing is the principal production system employed by the colonisers of rain forests and has been, and still is, encouraged by most state agencies for rural development and agrarian reform, even though scientific research has demonstrated clearly the failure of this system in most tropical ecosystems. Livestock production parameters in extensive grazing systems in tropical developing countries are notoriously poor. The average fertility rate rarely exceeds 50% and is often less; average stocking rates are less than one mature cattle unit per hectare; slaughter age for 450 kg live weight steers is more than 40 months; mortality rates frequently reach high figures in many regions due to contrasting food supply situations caused by long droughts and dry periods.

The recent evaluation of a dairying project in Costa Rica provides further confirmation of the unsustainability of tropical pasture-based livestock systems. Rain forest was cut down and burned in 1979-84 to establish *Brachiaria* pastures for family farm resettlement. In 1992, it was revealed that incomes had deteriorated, soil fertility had decreased, weeds had taken over from the *Brachiaria* and concentrate usage had increased. The authors concluded that tropical pasture milk Production was not sustainable and that research was needed to facilitate transition to other systems of land use.

When cattle grazing systems are the main activity of poor farmers, with insufficient capital and minimal access to credit, returns are usually inadequate to support the basic needs of the family. The consequence is that the land is sold usually to the rich landholders who, through economics of scale, can continue with the extensive grazing systems; the poor farmer turns once again to the forest, and so the destructive process continues. This cycle has been repeated through successive generations during the present century, the situation becoming increasingly severe since the areas cleared of vegetation are highly prone to crosive tendencies, especially the soils in the Andean and Amazonian regions.

The peasant farmer sector in Colombia (1–15 ha per farm) accounts for 70% of the rural population, supplies half the national food budget, yet occupies only 15% of the national territory. In contrast, extensive grazing systems occupy more than half of the agricultural area of Colombia, are owned by less than 5% of the rural population and still produce very little.

The contrast with Asian livestock production systems is interesting. In Vietnam, for example, erosion is not a serious problem and even the areas desiccated by defoliants during the war are regenerating vegetative cover. The reason for the environmentally-friendly role of livestock in Vietnam is that there is no recognised pasture-based beef industry. The role of cattle and of buffaloes is to supply the power needed by agriculture. They are therefore kept in the cropping areas and are fed almost exclusively on fibrous crop residues and by grazing on fallow and common lands.

Erosion

Africa's grazing systems are characterised by agro-pastoralism and transhumance. Such systems were apparently sustainable in times of low population density, with little pressure on the natural resource base and with opportunities to move from degraded lands to new territories or to adapt the pastoralist practice; but they have been destabilised by "development" practices, which have removed former "density-dependent" constraints, or added new constraints.

The impact of this destabilisation was clearly seen in the Dodoma region of Tanzania. Explosive growth of the population resulted in increasing areas of rangeland being diverted to cropping. At the same time, the livestock herds of the pastoralists were also increasing. The outcome was uncontrolled over-grazing of the non-cultivable areas, leading to severe land degradation, threatening total ecological collapse of the region. The seriousness of the situation resulted in the initiation of a far reaching and, in some respects, unique programme—The HADO Project (Hifadhi Ardhi Dodoma—Dodoma Region Soil Conservation Project).

The HADO project was started in 1973 and was initially concerned with arresting the accelerating land degradation occurring in parts of Dodoma Region through physical soil conservation measures. However, it quickly became apparent that the terraces, bunds, cut-off drains, etc. that had been constructed were not having the desired effect due to their destruction by grazing animals, and also due to uncontrolled water run-off from higher slopes denuded by over-grazing. As a result, a decision was taken in 1979 to close the most severely affected area of over 1,200 km^2—the so called Kondoa Eroded Area—to all livestock, which involved the eviction of over 85,000 cattle, goats, sheep and donkeys.

A review of the Kondoa area, 10 years after the decision to de-stock , showed that the regeneration of the vegetation, and the arrest of ecological degradation generally in these areas had been dramatic. Honouring the promise to the farmers that some form of livestock keeping would be allowed when the land had recovered, in 1990, the government, with help from SAREC and SIDA, introduced a zero grazing scheme for milk production with improved crossbred and local cattle. Results have surpassed expectations, with milk yields of up to 10 litres daily being achieved on locally available feed resources, and with major participation of women in the feeding and management of the cows and the use and sale of the milk.

Soil and water pollution

The problem of soil and water pollution has arisen due to excessive use of chemical fertilisers and insecticides in "green revolution" agriculture. Loss of soil organic matter, which increases the need for fertiliser inputs, through monoculture of exploitive crops such as cotton and cassava has been a contributory factor.

A related issue is the effect that excessive chemical fertiliser application and burning of crop residues has had on natural ecosystems. There is increasing evidence that high levels of nitrogen fertilisation decreases fixation of atmospheric nitrogen in the rhizosphere of, for example, sugar cane; and that it increases emissions of nitrous oxides and decreases oxidation of methane. By contrast, leaving post-harvest cane trash on the soil as a mulch, instead of burning it, increases sugar cane yields and soil fertility.

The integration of livestock with crops provides both nutrients for the plants and organic matter as an energy source for soil micro-organisms to aid soil fertility. On a specialised crop farm there may be little incentive for planting break crops of legume forages. But if livestock are present then such forages can be turned into income by feeding them to animals. Planting of multi-purpose nitrogen-fixing trees in association with cash crops, as in "alley farming" systems, is also more attractive to the farmer if some of the foliage can be used to give added value to livestock.

Loss of biodiversity

Genetic selection for livestock of ever-increasing productive potential has inevitably lead to decreased biodiversity at the animal level. Intensive feeding systems for monogastric animals, almost exclusively tied to use of cereal grains and soya bean meal, have encouraged replacement of indigenous ecosystems and local strains of cereals with 'more productive' hybrids. Emphasis on specialised grazing systems in the tropical savannahs has created vast expanses of pasture monocultures of *Brachiaria* spp. In both cases plant biodiversity has been reduced.

The positive side of increasing affluence is the opportunity to choose more on quality and less on price. In Colombia, eggs from scavenging 'local' poultry were preferred and brought higher prices that those from 'battery' birds. Meat from an indigenous pig breed had a better taste than that from imported 'improved' breeds and was preferred by local inhabitants in Guadeloupe. The meat from non-ruminant herbivores living in natural ecosystems is considered to be a delicacy in many tropical countries. The search for alternatives to cereal grains and protein-rich oilseed and animal by-product meals is already leading to the identification and promotion of a wide range of indigenous crop and water plants, trees and shrubs. Biodiversity will be enhanced by these practices which should be encouraged.

Human aspirations and the resource base

The economic strength and the standard of living of the industrial countries is directly linked with their consumption of fossil fuel. The aspiration of the less-developed countries is to follow a similar route. But reserves of fossil fuel are finite and have a lifetime measured in decades, not centuries. Hydro and nuclear power pose serious threats to biodiversity and to contamination with hazardous wastes, with fewer opportunities for employment.

The only sustainable solution is to promote life styles, and goods, which are derived from activities associated with the development and management of natural biologically-based resources. For the researcher in a tropical country, responding to this challenge should be a privilege and source of satisfaction, long since absent from the agenda of their colleagues in industrialised countries for whom agriculture is of declining importance.

The Need for Energy

The close link between livestock policies and fossil fuel use has been mentioned. Three examples put this in perspective. On the 30,000 ha sugar estate in the Dominican Republic, some 18,000 oxen haul the sugar cane from the fields to pickup points on a railway system leading to the sugar mill. This system is highly sustainable since the energy for the oxen is derived from the carbohydrates in the cane tops; nitrogen and minerals in the tops are returned to the soil in their excreta, since the animals eat and rest in the recently harvested areas.

By contrast, in Cuba, some 80% of the sugar cane is harvested mechanically by diesel-driven combines and loaded onto trucks which transport the stalks and attached trash to cleaning centres. Here, electrical power is used to blow off the trash and the stalks are elevated onto rail wagons or trucks for continued transport to the factory. This system is not sustainable. At the time of writing this manual, the problem of de-mechanisation of Cuba's agriculture was the subject of keen debate.

The example of Vietnam has already been mentioned, where agricultural power is supplied almost exclusively by buffaloes and oxen, and bicycles are the major means of personal transport. Vietnam's rating in terms of GNP may be one of the lowest in the world but, if it were assessed in terms of sustainable agriculture, it would be among the leaders. By contrast, in most tropical countries in Latin America, oxen have been replaced by tractors and forests are burned to develop pastures for beef cattle. These policies are highly unsustainable.

A specific problem of less-developed countries is the provision of domestic fuel for the more than 2,000 million families that use firewood from woods and forests for this purpose. In Colombia, it is estimated that 29% of the annual rate of deforestation is caused by domestic fuel wood consumption, half of which is in rural households, a third in poor urban communities, the remainder divided between charcoal production (9%) and rural handcrafts (11%).

The amount consumed varies from region to region, but it is calculated that a rural family of 8 persons uses 18 cubic metres of firewood annually solely for cooking. If this is purchased then it may cost up to US$70.00/tonne. When cut from the forest, it is estimated that 50 work days are expended annually for this purpose, with a value of US$147.00. The situation in much of Africa and in parts of Asia is similar.

Several solutions have been proposed which involve livestock. They are complementary and depend on natural and economic resources available and on cultural acceptance of the technology on offer:

— biogas digesters,

— establishing energy plantations, and

— use of crops that fractionate easily into "feed" and "fuel" components.

Biogas technology was first developed in India and China. The mixture of methane and carbon dioxide, produced by the anaerobic fermentation of livestock and human excreta, has found major uses in cooking. Biodigesters are intimately linked with livestock production, as they depend for substrate on the excreta of animals. Thus, use of this technology strengthens the arguments for partial or full confinement of animals, and thereby forms part of the strategy against uncontrolled grazing. The major constraint to the popularisation of biodigesters has been cost and availability of suitable materials for their construction. The recent development of low-cost biodigesters using standard polyethylene tubular film has had a major impact in Vietnam and

Cambodia where the even lower costs put the technology within reach of the majority of families.

Cereal crops are easily separated into grain and straw. The latter is burned for fuel on open fires in many developing countries. But increasingly in industrialised countries, especially those with strong legislation against uncontrolled burning, it is used as fuel in boilers designed for this purpose.

In Denmark whole villages are heated in the winter using this technology. In tropical countries, there are even greater opportunities for applying this principle. Production of sugar from sugar cane is one of the few agro-industrial activities which is self-sufficient in energy. In Vietnam, the growing of enough sugarcane to feed four pigs produced fibrous residues sufficient to cover half the fuel needs of a family of six.

The concept of the multi-purpose biomass refinery, in which the juice is the basis of animal feed or chemicals and the fibre is converted to synthesis gas to power gas turbines for electricity generation, promises to be a more viable—economically, sociologically and ecologically—than the single-purpose production of alcohol as in the Brazilian model. Multi-purpose trees can also be fractionated easily into feed and fuel and can thus be part of the same "integrated"model.

Energy plantations are important for the arid and semi-arid regions as they are complementary to pastoral-forestry schemes. Many species that can be used also fix atmospheric nitrogen and produce edible foliage and/or fruits. They include: *Acacia, Prosopis, Leucaena, Gliricidia, Guazuma, Inga, Albizia, Cassia, Pithecellobium* and *Alnus* spp. They may be sources of food, feed, fuel, timber and protection against erosion and desertification. These systems can also be the basis of biomass refineries as described above.

Thus the promotion of sustainable systems of agricultural production, in which livestock play a fundamental role, can also contribute to the solution of the domestic energy crisis. The use of multipurpose crop plants and trees, and the recycling of livestock excreta, provide not only much needed domestic fuel, but also control crosion, reduce contamination and act as sources of fertiliser.

Animal Welfare

Animal behavior studies were originally conceived as a means of exploiting livestock more efficiently through greater understanding of their habits and activities in different environments. The approach today is quite different. Behavior studies are done so as to develop less exploitive methods of animal production. The aim is to reduce stress to the animal and the attendant so that the quality of life of both is improved.

By contrast, the deliberate promotion of contentment through natural means can be reflected in higher productivity. The calf, lamb or kid that is suckled by its mother will grow faster, be healthier and have a better feed efficiency than if it receives its milk from a bucket. The dam will also respond to the more natural environment of having her offspring present at milking, and having it suck the residual milk from the udder. Milk yield will be higher and udder diseases less than if calves are weaned permanently soon after birth. Calves suckled naturally do not have the urge to suck the navels of their neighbours and thus can be managed in groups instead of being confined to individual pens. Sows fed fibrous feeds during gestation are less prone to develop anti-social behavior than when high nutrient density feeds are given. They can then be managed in groups rather than in separate individual stalls.

Stressful systems of livestock management, such as raising animals in cages and stalls, are already being legislated against in many countries in Europe. Practices such as debeaking of birds housed in cages, amputation of the tails of pigs and castration, reduce productivity and invite cannibalism.Embryo transplants have been heralded as a means of increasing beef cow profitability by inducing multiple births and thus raising prolificacy. However, this technology can result in a high degree of stress in both the cow and her attendant.

The long term effects are likely to be reduced lifetime fertility. Stimulation of cow milk yield by injecting recombinant growth hormone appears to reduce longevity and to increase stress through accelerated partitioning of nutrients from body tissues into milk. The welfare of these cows is certainly decreased and cannot be considered to be sustainable.

The direct economic cost of stressful systems of management will ultimately be reflected in the market place with premiums for products from contented and well cared-for animals and penalties for products of animals that are ill-treated.The transformation of both extensive cattle ranching and the highly intensive methods practiced in monogastric animal production, into more integrated systems in which the livestock play a catalytic and complementary role rather than being the primary goal, will bring with it related advantages in terms of animal welfare.

Food Quality and Security

In an increasing number of supermarkets and stores in the industrialised countries, premiums are paid for food produced in "environmentally-friendly" farming systems. Crops that are grown according to "organic" farming principles are in this category; as are animal products derived from such cropping systems. The ban on imports to EU countries of beef from cattle treated with synthetic hormones shows how this concern for more natural food translates into economic criteria.

Constraints in Sustainable Livestock-based Agriculture

The issue here is that, in contrast to crop production, livestock systems in tropical developing countries have been highly influenced by practices developed in the industrialised countries, most of which are in temperate climatic zones. For example, most "modern" methods of pig, poultry and dairy production in tropical countries are almost exact copies of those practiced in industrialised countries, using the same germ plasm and feed resources. The term 'assembled' is often used to describe the products derived from such systems to emphasize their dependency on imported inputs.

Such practices have been justified by the need to respond to the aspirations inherent in a 'better standard of living' through increased consumption of food of animal origin. In fact, they exacerbate the basic problems since they result in:

— Minimum employment opportunities.

— An increase in the foreign exchange deficit, due to high imports.

— More pollution, as usually the animal population in such units is high and there are no associated crops for recycling the excreta.

— Impoverishment of the small scale farm family, which cannot compete in the purchase of the required inputs and may not have the skills for the more sophisticated management that is required.

Countries such as Nigeria and Venezuela, which built up sophisticated intensive animal industries in times of high oil revenues, found that these were not sustainable when oil prices fell and agricultural subsidies had to be reduced. Indicators of sustainability are derived from measurements which describe the effect of the system on the sustainability of the resource. While this topic is presently the subject of much discussion, the following parameters are proposed as criteria on which the sustainability of resource utilisation can be measured. The items include:

- — Total biomass yield.
- — Soil organic matter content.
- — Soil pH.
- — Soil content of P,N,K, and Ca.
- — Degree of diversity of animal genetic resources and their use at the level of small scale users.
- — Water quality.
- — Production and use of renewable energy.
- — Use of fossil energy.
- — Energy balances.
- — Diversity in fauna and flora at plant and soil level.
- — Greenhouse gas emissions and sinks.
- — Employment generation.
- — Involvement of women and children.
- — Food security.
- — Catalytic role of livestock in the integration of crops, livestock and forestry.
- — Protection against crosion and desertification.

Bibliography

Alam J., *Livestock Resources in Bangladesh,* University Press Ltd., Dhaka, Bangladesh. 1995.

______________., Rahman A., Sayeed M.A. and Hossain M., "A study on livestock credit in Bangladesh", *Bangladesh Journal of Livestock Research,* 1993.

Anon, "Livestock development in India: Issues and an agenda for future research", *Indian Journal of Agricultural Economics,* 1997.

Devendra C., "Crop residues for feeding animals in Asia: Technology development and adoption in crop–livestock systems", In: Renard C. (ed), *Crop Residues in Sustainable Mixed Crop/Livestock Farming Systems,* CAB (Commonwealth Agricultural Bureaux) International, Wallingford, UK, 1997.

______________., Thomas D., Jabbar M.A. and Kudo H., "Improvement of Livestock Production in Crop–Animal Systems in Rainfed Agro-ecolological Zones of South-East Asia", *ILRI (International Livestock Research Institute*), Nairobi, Kenya, 1997.

Doornbos M. and Gertsch L., "Sustainability, technology and corporate interest: Resource strategies in India's modern dairy sector", *Journal of Development Studies,* 1994.

______________., van Dorsten F., Mitra M. and Terhal P., *Dairy Aid Development: India's Operation Flood,* Sage Publications, New Delhi, India, 1990.

Gardiner P. and Devendra C. (eds), "Global Agenda for Livestock Research", *Proceedings of a Consultation,* ILRI, Nairobi, Kenya, 18–20 January 1995.

George P.S., "Dairying and livestock economy of India—a review", *Indian Journal of Agricultural Economics,* 1996.

de Haan C., Steinfeld H. and Blackburn H., "Livestock and the Environment: Finding a Balance", *European Commission Directorate-General for Development,* Brussels, Belgium, 1997.

Huq M.F., "Livestock development in Bangladesh for rural poverty alleviation", In: Gootjes C.P., den Herteg G., de Jong R. and Nell A.J. (eds), *Proceedings of the International Workshop on Livestock Production in Rural Development—Development of Livestock Policies,* Wageningen, The Netherlands, 21–31 January 1992.

Lensch J., "The Indian dairy industry as part of a sociologic, ecologic and economic system", Praktische Tierarzt, Germany, 1990.

Reynolds L., Thomas D., Hendy C.R. and O'Reilly C., "Environmental Impact Assessment of Livestock Production Systems in the Tropical Highlands", NRI (Natural Resources Institute), Chatham Maritime, UK, 1995.

Sastry N.S.R., "Livestock Sector of India: Regional Aspects", International Book Distributing Co, New Delhi, India, 1995.

World Bank, "India Livestock Sector Review: Enhancing Growth and Development", *Report 14522-IN,* The World Bank, Washington, DC, USA, 1996.

Index